Behavioral Neuroscience

Edited by Sara Palermo and Rosalba Morese

Published in London, United Kingdom

IntechOpen

Supporting open minds since 2005

Behavioral Neuroscience
http://dx.doi.org/10.5772/intechopen.73402
Edited by Sara Palermo and Rosalba Morese

Contributors
Valentina Ignatova, Maria Uscinska, Silvio Bellino, Estela Castilla-Ortega, Patricia Sampedro-Piquero, Luis J. Santín, John Stewart, Olga Penagarikano, Marta Fernández, Teresa Sierra-Arregui, Ita Puusepp, Tuisku Tammi, Minna Huotilainen, Teija Kujala, Elina Kuusisto, Sonja Laine, Kirsi Tirri, Götz Egloff, Dragana Djordjevic, Sara Palermo, Rosalba Morese

First published in London, United Kingdom, 2019 by IntechOpen
IntechOpen is the global imprint of INTECHOPEN LIMITED, registered in England and Wales, registration number: 11086078, 7th floor, 10 Lower Thames Street, London, EC3R 6AF, United Kingdom
Printed in Croatia

British Library Cataloguing-in-Publication Data
A catalogue record for this book is available from the British Library

Additional hard and PDF copies can be obtained from orders@intechopen.com

Behavioral Neuroscience
Edited by Sara Palermo and Rosalba Morese
p. cm.
Print ISBN 978-1-78984-051-3
Online ISBN 978-1-78984-052-0
eBook (PDF) ISBN 978-1-78985-185-4

For EU product safety concerns:
IN TECH d.o.o., Prolaz Marije Krucifikse Kozulić 3, 51000 Rijeka, Croatia, info@intechopen.com or visit our website at intechopen.com.

Meet the editors

Sara Palermo has an MSc in Clinical and Community Psychology and a PhD in Experimental Neuroscience. Currently, she is a postdoc research fellow at the University of Turin (Italy). She is a research member of the Center for the Study of Movement Disorders and the Placebo Responses Mapping Group at the Department of Neuroscience. She is also a research member of the Neuropsychology of Cognitive Impairment and CNS Degenerative Diseases Group at the Department of Psychology.
Sara Palermo is an ordinary member of the Italian Society of Neuropsychology, the Italian Association of Psychogeriatrics, the Italian autonomous association adhering to SIN for dementias, and the International Society for Interdisciplinary Placebo Studies. Importantly, she is involved in the European Innovation Partnership on Active and Healthy Aging.

Rosalba Morese holds a Bachelor's degree in Psychology at the University of Parma and a PhD in Neuroscience at the University of Turin to develop new techniques and approaches in cognitive and social neuroscience.
She is an expert in cognitive and social neuroscience, neuroeconomics, psychophysiology, and psychology. She performs neuroimaging studies in social contexts to investigate neural correlates involved during social interactions, such as social exclusion, social support, empathy, communicative intention, and social decision making. She is currently working at Università della Svizzera italiana, Switzerland.

Contents

Preface

What do we mean by "behavioral neuroscience?"

This book aims at providing an overview of behavioral neuroscience and deepening neuronal mechanisms and brain circuits that regulate the fundamental aspects of human behavior, such as cognitive and emotional functions. It is intended to give the reader the most up-to-date vision of how the interaction between biological mechanisms and neurocognitive processes leads to complex and highly organized behaviors.

The authors offer original contributions to develop new perspectives in behavioral neuroscience thanks to the originality of their ideas, theories, research, scientific results, and discussions.

The first part of the book deals with the definition of behavioral neuroscience and the presentation of the framework in which it is located.

The introductory chapter emphasizes how behavioral neuroscience concerns not only the biological bases of behavior but also the more complex phenomena of mind and brain. Indeed, the great challenge of neuroscience is to understand behavior and thought. One of the main topics of discussion in the twentieth century was whether mental activities are functions different from brain activities or if they also represent functional expressions of the brain neurons. Only in the twenty-first century do we begin to have information of fundamental importance on the nature of highly organized and complex mental processes such as consciousness, will, and social cognition. Today we still find ourselves having to answer this question: *Are we our brain?*

The second chapter opens the book on the interesting "neurophenomenological" perspective to establishing correlations between descriptions of lived experience and brain states. The question we want to answer is: *How can a neuronal state be a state of consciousness?*

The third chapter describes the contribution of intestinal microorganisms for modulation of many systems and human behavior. The author thus underlines well how exploring the interaction of gut microbes and human brain will not only allow us to better understand the pathogenesis of neuropsychiatric disorders, but will also provide us with new opportunities for the design of novel immuno- or microbe-based therapies.

The second part of the book deals with the contributions of behavioral neuroscience in the field of child neuropsychiatry.

The first chapter of this section is dedicated to social factors and early psychosocial interventions that can modify (in a positive way) the trajectories of the mind–brain relationship. The second chapter outlines the role of mindsets and failure in explaining learning differences among students. An experimental study in the

Finnish elementary school context is the starting point for a constructive discussion in the light of earlier neuroscientific research related to mindsets, including limitations and suggestions for future research in the field. The last chapter of the section opens to the topic of neuropsychiatric pathology in children. The association between cerebellar neuroanatomical alterations and autism is analyzed, opening a new avenue for further research.

The third part of the book deals with the contributions of behavioral neuroscience in the field of adult neuropsychiatry.

The first chapter of this section is aimed at an in-depth examination of treatment-induced brain plasticity in psychiatric disorders. Although neural substrates of symptoms expression have been studied extensively, neural mechanisms mediating post-treatment amelioration of symptoms and brain networks functionality remain poorly characterized. We now have to rethink mental disorders, re-evaluating the treatment possibilities given by learning and brain plasticity.

The last chapter reviews the altered brain structure and function associated with drug addiction, with a focus on brain regions involved in learning and motivated behavior. Particular attention is paid to the consequences of reduced ability to experience rewards and emotional dysregulation, leading to persistent memories of pleasure related to drugs responsible for the onset of the harmful "addiction cycle."

It is a small book with great content. The book offers an excellent synopsis of perspectives, methods, empirical evidences, and international references. Therefore, it represents an extraordinary opportunity to target neuroscientific hot topics and outline new horizons in the study of the relationship between brain and behavior.

> *"Neuroscience is by far the most exciting branch of science because the brain is the most fascinating object in the universe. Every human brain is different—the brain makes each human unique and defines who he or she is."*
>
> *Stanley B. Prusiner*

Sara Palermo, PhD
Department of Psychology,
University of Turin,
Italy

Rosalba Morese, PhD
Faculty of Communication Sciences,
Università della Svizzera Italiana,
Switzerland

Section 1

Definition and Operational Framework of the 21st Century

Chapter 1

Introductory Chapter: Neuroscience *Wants* Behavior

Sara Palermo and Rosalba Morese

1. Introduction

What do we mean by "behavioral neuroscience"? This is the branch of neuroscience developed from Wilhelm Wundt's and William James's physiological psychology and addressed to the study of the "the neural and biological bases of behavior, including effects of lesions and electrical stimulation, recording of electrical activity, genetic factors, hormonal influences, neurotransmitter and chemical factors, neuroanatomical substrates, effects of drugs, developmental processes, and environmental factors" [1].

Historically, neuroscience is born with the identification of the *neuron* as an autonomous and functionally independent cellular unit of the nervous system. The studies carried out to define the properties of the neuron have benefited from the progress made in various disciplines, in particular using methods to measure ionic and molecular displacements at the subcellular level and—thanks to the original psychopharmacology, psychophysiological, and neuroimaging approaches—the progress made in the knowledge of integrated systems at the base of the behavioral variations of the individual [2]. In the beginning, neurotransmitters such as acetylcholine, 5-hydroxytryptamine, and GABA have been discovered, and the structural aspects of membrane receptors for different molecules with neurotransmitter functions have been analyzed. Subsequently and of particular interest was the identification of endorphins and their receptors on nerve cells [2]. With the identification and study of endorphins, a new approach to the analysis of substances that perform a modulating function on the genesis and transmission of nerve impulses has been developed. As part of the research on cellular differentiation by chemical substances, molecules have been identified that play a fundamental role on the growth and tropism of the nerve cell [2]. The prototype of these substances is the nerve growth factor isolated in the early 1950s by the Italian neuroscientist Levi-Montalcini [3].

In short, neuroscience initially contributed to defining the functioning of the neuron and the role of neurotransmitters, of neuromodulating molecules, and of those with trophic action.

In parallel, the biomedical approach has allowed the use of various investigation techniques to explore the anatomic-functional structure of the nervous system as an integrated unit, both in normal and pathological conditions. In this sense, the progress made in neuroradiology and neuroimaging must be seen. The neuroscience approach has therefore extended to the description of molecules able to control the genesis of some brain proteins (the so-called genetic engineering) [2]. In particular, molecular biology has allowed us to study amino acid sequences of peptides that seem to play a physiological or pathological role, in relation to the different conditions of isolation and characterization [2].

To date, behavioral neuroscience also includes psychoneuropharmacology studies, which have analyzed the complex interactions between substrates of the central nervous system, the distribution of various molecules, and the state of brain functioning [2]. The result is practical acquisitions, which are extremely important in the synthesis of psychotropic drugs widely used in the therapy of neurotic and psychotic states.

Behavioral neuroscience concerns not only with the biological bases of behavior. It concerns with the more complex phenomena of the mind and brain.

2. From brain activities to mental activities

Not only *bío* but also *psyche*: the great challenge of neuroscience is to understand behavior and thought.

Although humans have wondered about the control of behavior for thousands of years, only fairly recently has a mechanistic view of the brain taken hold [4, 5]. The concept of *localization of function* was an important milestone for behavioral neuroscience. Today we know that the contemporaneous functional modulation of different cerebral areas varies in a predictable way depending on what a subject is doing. Thanks to modern neuroimaging and a more carefully validated understanding of human cognition, a detailed view of the brain organization is now emerging. Modular systems are outdated; the network approach is the current one [2].

One of the main topics of discussion in the twentieth century was whether *mental activities*—such as thought, emotions, self-awareness, and will—are functions different from *brain activities*, such as the movement of a limb, the perception of a color, etc., or if they also represent functional expressions of the brain neurons. Mental and cerebral activities would seem to be the unique and indivisible expression of the activities of the neuronal and glial elements that make up the brain organ. Although the expression is different in quality and in the ways in which it manifests itself, both activities are due to a single mechanism by which neurons communicate with each other and with the rest of the body [2]. The neural circuits and their realization are encoded in the animal genome, while the environmental stimuli play a fundamental role for the definitive realization of the synaptic connections.

Neurons, organized in ganglia, complex structures, and networks, process nerve impulses, memorize them, and emit behavioral responses. It is probable that once we fully understand these first two levels—that of the functions of the individual neurons and that of the activities of the neural networks—we will arrive at the elucidation of the type of circuits (or nervous activities) with which subjects are able to decide a specific motor act or a reminiscent act and the mechanisms by which the brain, at the same time as processing sensory inputs, makes subjects aware of all these operations.

Only today, we begin to have information of fundamental importance on the nature of mental processes such as consciousness, will, social cognition, and enormously complex problems that constitute the core of the third level of brain functions.

3. Brain-behavior relationship

Disorders of the brain and nervous system (such as Alzheimer's disease, Parkinson's disease, stroke, and traumatic brain injuries) highlight the importance of the biological bases of behavior. The *brain-behavior relationship* seems to be the descendant of the Cartesian mind-body dualism, where the brain is the biological component and behavior the psychological one.

Despite the passing of the centuries, the body-mind dualism continues to be an unresolved problem in this day and age. At the beginning of the history of neuroscience, the mind and brain were kept apart as if they were separate and distinct concepts. Nevertheless, the notion that the brain and behavior function separately turns out to be an impediment to scientific progress, since they are related in a more complex way than one might imagine. Indeed, the brain-behavior relationship is modulated by different factors: the environment, sociocultural and historical aspects, phylogeny, genetics, and ontogeny.

Today we still find ourselves having to answer this question: *Are we our brain?*

As recently suggested by Dede and collaborators, "the development of brain-behavior relationship depended thereafter on interdisciplinary collaboration, and scientists' ability to formulate new experimental questions and designs, but mainly on the methods devised for studying both parts of this dipole" ([6], p. 12). Today, behavioral neuroscientists balance three general research perspectives in designing their research [6]:

- Correlation: body and behavioral measures covary.
- Somatic intervention: manipulating the body may affect behavior.
- Behavioral intervention: experience affects the brain.

Indeed, behavioral neuroscience research is now conducted at a level of analysis ranging from molecular events to the functioning of the entire brain and complex social situations.

4. The nature and purpose of the book

Revising the story of brain-behavior relationship research since its beginning, a treasure of information about how the brain works seems to have been discovered. Nevertheless, neuroscientists' research will continue until no more questions can arise [7]. Indeed, the benefits of the correlation-somatic-behavioral research approach [6] are greatly enhanced by the combined use of new technologies: neuroimaging, ICT-IoT assessment methods, and machine learning. It can be assumed that computerized functional localization is now a reality and that the brain-behavior relationship has already moved on to the next stage of its development (what we have just named the third level of brain functions). Considering the above, this volume aims at providing an overview of behavioral neuroscience and deepening neuronal mechanisms and brain circuits that regulate the fundamental aspects of human behavior, such as cognitive and emotional functions. It is intended to give the reader with the most up-to-date vision on how the interaction between biological mechanisms and neuropsychological processes leads to complex and highly organized behaviors. The approach is multidisciplinary, and various levels of investigation are represented.

Author details

Sara Palermo[1*] and Rosalba Morese[2]

1 Department of Psychology, University of Turin, Turin, Italy

2 Department of Psychology, Center for Cognitive Science, University of Turin, Turin, Italy

*Address all correspondence to: sara.palermo@unito.it

References

[1] Thompson RF. In: Smelser N, Baltes P, editors. International Encyclopedia of the Social and Behavioral Sciences. Oxford, UK: Pergamon (Elsevier); 2001. ISBN 9780080430768

[2] Kandel ER, Schwartz JH, Jessell TM, Siegelbaum SA, Hudspeth AJ. Principles of Neural Science. 5th ed. New York: McGraw-Hill; 2012. ISBN 0-07-139011-1

[3] Cohen S, Levi-Montalcini R, Hamburger V. A nerve growth-stimulating factor isolated from sarcom as 37 and 180. Proceedings of the National Academy of Sciences. 1954;**40**(10):1014-1018

[4] Lezak MD. Domains of behavior from a neuropsychological perspective: The whole story. Integrative views of motivation, cognition, and emotion. In: Nebraska Symposium on Motivation. Vol. 41. Lincoln: University of Nebraska Press; 1994. pp. 23-55

[5] Lezak MD. Neuropsychological Assessment. 5th ed. Oxford, UK: Oxford University Press Inc.; 2012. ISBN: 9780195395525

[6] Dede E, Zalonis I, Gatzonis S, et al. From discovering to better understanding the relationship between brain and behavior. Integrative Neuroscience Research. 2017;**1**(1):5-16

[7] Reitan RM. Certain differential effects of left and right cerebral lesions in human adults. Journal of Comparative and Physiological Psychology. 1955;**48**:474-477

Chapter 2

Neurophenomenology, Enaction, and Autopoïesis

John Stewart

Abstract

The project of neurophenomenology, initiated by Francisco Varela, aims at establishing correlations between descriptions of lived experience (as elicited by the explicitation interview technique) and brain states (as measured with increasing precision and detail by the new brain imaging techniques). However, on their own, such correlation aggravates rather than solve Chalmers' "hard problem"–how can a neuronal state *be* a state of consciousness? The question that arises is thus how to interpret such correlations. I will argue that this requires putting the brain in the body of an animal living in the world. Epistemologically, this amounts to putting neuroscience in the context of cognitive science (Varela's concept of enaction) and cognitive science in the context of biology (Maturana and Varela's concept of autopoïesis).

Keywords: neurophenomenology, enaction, autopoïesis, conscious experience, cognition, life, Varela

1. Introduction

In 1995, Chalmers [1] identified what he called the "hard problem" of consciousness. The question is, how can a purely physicochemical event such as a set of action potentials in a set of brain cells (and why brains, rather than the heart or the gut?!) actually *be* a subjective, conscious experience. Chalmers wrote: "Why is it that when our cognitive systems engage in visual and auditory information processing, we have visual or auditory experience: the quality of deep blue, the sensation of middle C? ... Why should physical processing give rise to a rich inner life at all? It seems objectively unreasonable that it should, and yet it does." Over 20 years later, we are hardly any nearer a solution; the problem remains entire.

In Section 2, I will present the "neurophenomenology" proposed by Varela [2], which provides empirical evidence on the neural correlates of consciousness. While entirely valid and indeed intriguing, this work does not solve the "hard problem"; I will argue that in a way it makes matters even worse, because the correlations it reveals are not explanatory, but they require explanation-although in the conclusion, I will suggest that there is actually more to it than this.

In Section 3, I will present the concept of "enaction" [3]. In order for conscious experience to be possible, there must be something to be conscious *of*: "what is it like? ...", the question of *qualia.* Enaction is the process, whereby a living organism brings about or enacts its own characteristic lived world of experience. The classic example is the "world of the tick" as evoked by von Uexküll [4], originally in 1902, who provided a compelling view of "what it is like to be a tick." An additional feature is that whereas the original, prototypical example of enaction is the lived world of an animal, and this concept also applies to humans. And more specifically, this means that each and every one of the humans *enacts* our own lived world, every minute of every day of our lives. This introduces a note of first-person subjectivity, which contrasts with the third-person objectivity more usual in scientific discourse.

In the previous paragraph, I said that enaction is the process, whereby a *living organism* brings forth its lived world of experience. This raises the question as to whether being alive may not be a prerequisite for experiencing consciousness. There are two aspects to this: on the one hand, a "brain in a vat" would arguably not be properly conscious; on the other, it is only in science fiction that mechanical robots with computerized "brains" can be conscious. In order to deepen our understanding of what it takes to be fully alive, in Section 4, we will look at *autopoïesis,* the process whereby living organisms continually *fabricate themselves*. This is something that every living organism does, even a lowly bacterium; but a brain in a vat, just by itself without being part of an organism, does not; neither does any man-made machine, not even the most powerful computer imaginable with or without a sophisticated connectionist architecture.

In the box below, I provide definitions of some major concepts in this field.

Definitions of major concepts

Enaction. *Enaction* is the process, whereby a living organism *brings forth* its own specific "lived world." The prototype example is the "world of the tick," as described by von Uexküll and detailed in the text. It is to be noted that "enaction" also applies to human beings, including each of us, ourselves, which introduces a dimension of first-person subjectivity unusual in scientific discourse.

Autopoïesis. *Autopoïesis* is the process, whereby living organisms continually *produce themselves.* They are thus *processes* rather than *things*; they only become "things" again when they are dead. This feature differentiates the humblest living organism, even a bacterium, from the most sophisticated robot with a computer "brain," since these robots do *not* produce themselves, they are made and repaired from the outside, by human engineers; in other words, they are *heteropoïetic.*

Dissipative structure. In thermodynamically *open systems* (qv), there can be spontaneous generation of dynamic *dissipative structures,* dynamic forms which are *par excellence* processes rather than things. Mundane examples are tornadoes or a whirlpool in a river. Living organisms belong to the class of dissipative structures. However, whereas common examples such as whirlpools are essentially ephemeral, lasting only as long as special external conditions over which they have no controls maintained; living organisms are potentially immortal, having the capacity to perpetuate themselves indefinitely (see *autopoïesis*).

Thermodynamic systems: open and closed. In classical thermodynamics, there are two sorts of system. In a *closed* system, there are no exchanges with anything external to the system; it is in such systems that "entropy" (i.e. disorder) can be properly calculated and it demonstrably increases monotonically. In *open* systems, by contrast, there is a continual flow of energy and matter through the system. In this case, "entropy" cannot be properly calculated, and so no longer increases. In such systems, dynamic *dissipative structures* (qv) can arise spontaneously, and maintain themselves indefinitely.

Finally, in conclusion, I will attempt to draw these various threads together.

2. Neurophenomenology

A notable attempt to solve the "hard problem"—or rather, to dissolve it away—is the neurophenomenology proposed by Varela [2]. There are two strands to neurophenomenology. On one hand, there are the modern techniques of brain imaging, which provide rich empirical data—fine-grained, precisely localized, in real time—on cerebral activity. On the other hand, there is a serious attempt to obtain valid descriptions of first-person subject consciousness. It is notoriously difficult to obtain such descriptions; spontaneous introspection is disconcertingly unreliable. In a well-known experiment, repeated by various authors, two similar subjects are

Attention. Three attentional networks contribute to distinguishing conscious from nonconscious cognitive events: orienting to sensory stimulation, activating patterns from memory, and maintaining an alert state.
Present-time consciousness. Phenomenological studies reveal a basic three-part structure of the present, with threads into past and future horizons. These structural invariants link naturally to observations in cognitive neuroscience, which show that there is a minimal time required for the emergence of neural events that correlate to a cognitive event. This noncompressible time framework can be analyzed as a manifestation of the long-range neuronal integration in the brain linked to a widespread synchrony.
Body image and voluntary motion. The nature of the will as expressed in the initiation of a voluntary action is inseparable from consciousness and its examination. Recent studies give an important role to neural correlates, which precede and prepare voluntary action, and the role of imagination in the constitution of a voluntary act.
Perceptual filling-in. This phenomenon involves the spontaneous completion of a percept, so that the appearance (i.e., a visual contour) is distinct from the physical correlate (incomplete borders, as in illusory contours). The neuronal data on filling-in correlate well with the phenomenological observation: there is an important difference between "seeing as" (visual appearance) and "seeing what" (a visual judgment).
Fringe and center. In traditional phenomenology, there is a two-part structure of the visual field between a "center" and a "fringe." This correlates well with the structure of the retinal, between "foveal" area with a high density of retinal cells and a "peripheral" area with much lower density.
Emotion. In recent years, there have been significant advances in identifying the brain correlates of emotions. Evidence points to the importance of specific neuronal structures such as the amygdala, the lateralization of the processes involved, and on the role of arousal in emotional memory.

Table 1.
Basic brain mechanisms for consciousness.

shown, but slightly with different portraits, and were asked to designate the one they prefer. They are then shown a slightly enlarged version of the portrait, *other* portrait, and asked to explain why they chose it. Blithely unaware of the subterfuge, the subjects dutifully spin a yarn explaining why they preferred the portrait they had *not* chosen. At a more basic level, when asked to describe their experience of a visual perception, naïve subjects base themselves on what they know, or think they know, about the object in question (an apple, a house, a dog…). It was in order to counter this tendency that Husserl [5], the founding father of phenomenology, developed the practice of "eidetic reduction," that is, putting aside or bracketing out one's knowledge of the object, in order to focus on the actual experience itself. This is not easy, but possible if due care is taken. In Varela's own work, he employed an "interview method" to obtain reliable descriptions of subjective experience [6].

With these methods and also drawing on the literature, Varela obtained some convincing results. The major results are detailed in **Table 1**.

So where does this leave us? These neural correlates of consciousness are, in their own way, impressive indeed. However, with respect to the "hard problem," there is a sense in which we are worse off than ever. The point is that these correlates are just that; they do not actually *explain* anything, and indeed they themselves require explanation. There is actually something more to Varela's neurophenomenology than this, and I will return to this point in conclusion. However for the moment, we will now take a look at what might lie behind these correlations.

3. Enaction

To start with, it will be salutary to go back to basics and look at what it is that neurons actually *do*. Physiologically, their primary role is to connect sensory inputs to motor outputs. They are well suited to this task, since their basic mode of action is the action potential; and the cells of both sensory organs (eyes, ears, noses, tactile receptors, etc.) and effector organs (principally muscles) also function with action potentials. It is thus quite convenient to connect receptor organs to neurons, and neurons to effector organs, by means of synapses. The point here is that by this means, the actual connections between sensory organs and effector organs is quite flexible: any sensory organs can be connected to any effector organ, and the connection can be excitatory or inhibitory depending on what is appropriate in terms of the sensory-motor dynamics thus set up. In order to appreciate the ecological significance of this, the time has come to introduce the notion of *enaction*, the process whereby a living organism enacts, or brings about, its own characteristic "lived world."

The prototype example is the "world of the tick" as described by von Uexküll [4]. This lived world is enacted by three simple sensation-action cycles. (i) The female tick climbs to the end of the branch of a bush, and ... waits, maybe for weeks. If she senses an odor of butyric acid, she lets herself fall. (ii) If she falls on a hairy surface, she crawls until she finds a smooth area (if not, she climbs back up onto the bush and starts over). (iii) She then sticks her proboscis through the surface, and if she finds a liquid at 37°C underneath, she sucks to satiety. This makes sense when we know that (in context) butyric acid is secreted by the sweat glands of mammals; the hairy surface will then be the fur of the mammal; and the liquid at 37°C will be the blood of the mammal (which the tick needs to feed her eggs; she then fertilizes the eggs from a store of sperm, after which she lays the eggs and her life cycle is complete). Of course, the tick does not know *that,* as such; indeed, she is barely conscious. But the important point for us here is that these simple sensory motor dynamics are set up by the wiring of the tick's simple peripheral nervous system. Note that this would not work if activation of the butyric acid sensors did anything other than triggering the motor action of letting herself fall; if the tactile sensors of a hairy *versus* smooth surface triggered anything other than crawling to a smooth surface, and sticking the proboscis through the surface; and if the temperature sensors of liquid at 37°C did not trigger the action of sucking. The example is magnificent; in this way, the tick, a tiny animal, which can only crawl slowly, manages the feat of catching a mammal far bigger and faster than herself, and not only that but getting to suck its blood. Thus, although the tick is barely conscious, the wiring of her nervous system is instrumental in setting up the meaningful lived world that she is minimally conscious *of.* This is indeed a basic point; Husserl noted that "consciousness" is always consciousness *of*, consciousness is always "intentional," aiming *at* something meaningful. So what we see here is that way before we get to the higher level consciousness experienced by humans, the nervous system is fundamental to setting up the whole situation.

Before moving on, there is another aspect to enaction; that is, as conscious living beings, each and every one of us *enacts* our own lived world, every minute of every day. This introduces a note of first-person subjectivity that is unusual in scientific discourse; I will return to this point also in the conclusion.

4. Autopoïesis

I now address the point that consciousness would seem to be a feature reserved to living organisms; it is only in science fiction that computers and computerized robots are conscious (the film "Space Odyssey 2001" with the computer "Hal," and the film "Her" in which the hero falls in love with his "operating system"). This raises the question as to what it is about living organisms that makes them candidates for consciousness. After all, computerized robots can perfectly have artificial nervous systems that set up their sensory motor dynamics. One lead is that all living organisms, even the simplest, have the property that they are *autopoïetic*, that is, they are pure processes that are continually "making themselves"; and this is something that even the most powerful computer, or the most sophisticated robot with a "brain" having a connectionist architecture, even attempts to do. Living organisms are thermodynamically open systems, with a continuous flow of energy and matter, which are essential to their existence (a living organism only becomes a "thing" again when it is a dead corpse). In this respect, living organisms are similar to other "dissipative structures" such as cyclones, or eddy currents in a river; with the difference that they actively promote the conditions that will enable them to perdure, which cyclones and rivers do not. It is in order to do this that they engage actively with their environment, in the course of which they enact their lived worlds. So maybe this is why a "brain in a vat" would arguable not be conscious; it is a brain *in a living animal, in the world*, which appears as the seat of consciousness. Exactly *why* consciousness is reserved to living organisms is not entirely clear, and worthy of deeper study.

Ironically, it is not the most highly evolved forms of human consciousness that are the most mysterious; Jaynes [7] has provided a highly suggestive account of how sophisticated reflexive consciousness might have arisen through the "breakdown of the bicameral mind". The "bicameral mind" in question is the sort of trance in which prophets (Old Testament) and heroes (Odyssey) "heard voices" telling them what to do. Nowadays, we call this "auditory hallucination" and shut up those concerned in a psychiatric asylum, whereas at that time, they were valued members of their communities. However, that may be, "hallucinations" are clearly the result of cerebral activity, as are dreams to which they are closely related. No; what is most mysterious is the basic form of "animal consciousness," which seems not only to be all-or-nothing but also to increase gradually along the evolutionary scale from fish to reptiles to mammals. This *correlates* with brain size but once again, correlation is not cause; a correlation is not an explanation, but rather a feature, which remains *to be explained.* Discussions of a scientific approach to consciousness are sometimes compared to the question of vitalism; living organisms were long considered as mysterious. Optimistic objectivists now consider that the advances in biology have rendered vitalism a problem of the past; with the suggestion that if we just continue with enough good natural science, the same will happen for the question of consciousness. I would like to suggest that things may not be so entirely straightforward.

5. Conclusion

In conclusion, I would like to return to some issues that I raised earlier. As we have seen, Varela's "neurophenomenology" highlights some of the neural correlates of consciousness. But as I said, if this were all, it would aggravate rather than solving Chalmers' "hard problem." But in fact it is not all. Varela's main point—which may be well missed by those reading it as one more academic text—is that if we really wish to address the phenomenon of consciousness, we must be prepared to take into account the fact that it is a phenomenon that is only properly instantiated in the form of *first-person subjectivity.* And this means that if we wish to seriously engage with it, we cannot escape the necessity of ourselves bringing into play our *own* individual consciousness. Varela writes:

> *"One must take seriously the double challenge my proposal represents. First, it demands a re-learning and a mastery of the skill of phenomenological description. Second: a call for transforming the style and values of the research community itself.*

> *To the long-standing tradition of objectivist science this sounds anathema, and it is. But this is not a betrayal of science: it is a necessary extension and complement. Science and experience constrain and modify each other as in a dance. This is where the potential for transformation lies. It is also the key for the difficulties this position has found within the scientific community. It requires us to leave behind a certain image of how science is done, and to question a style of training in science which is part of the very fabric of our cultural identity".*

The consequence is that with respect to the "hard problem," the nature of "hard" becomes reframed in two senses:

1. It is hard work to train and stabilize new methods to explore experience.

2. It is hard to change the habits of science in order for it to accept that new tools are needed for the transformation of what it means to conduct research on mind and for the training of succeeding generations.

Interestingly enough, a similar theme comes up with respect to the concept of enaction. As I have already indicated, there is potentially an existential dimension to enaction; scientists are after all human beings, as thus like all living organisms *enact* their lived world every minute of every day. Taking this personally, it means that I am actually *responsible* for the quality of what my enacted world leads me to experience. We are on dangerous ground here for "normal science": science is supposed to aim at objectivity; and it is very widely supposed that attaining objectivity requires the elimination of subjectivity. But subjectivity, if it is assumed as such, is neither more nor less than first-person experience as lived from the inside; and we have just seen that precisely, which is at the core of enaction. In other words, enaction, if it is taken seriously in what I personally see as its core, poses a manifest threat to our normal functioning as scientists.

Of course what happens is that any "normal" scientist will seek to defuse this threat – both individually and as a community. I do hope I am not being pretentious or disdainful here; I think I understand too well what is going on, because I know the cost. But still I do want to stand my ground, to stand up, and be counted. I maintain that if enaction is defused, it is betrayed. I propose that we take a closer look at this.

One of the main ways—certainly not the only one—that enaction defused is by converting it into a much safer research program of what has been called "4E cognition" [8]. The "4Es" are: embodied, embedded, extended, and enacted cognition. This is a smart move (if one is indeed trying to defuse enaction so as to get back into the comfort zone of "normal" science), for the following reason. Varela himself envisaged enaction as the framework for a possible paradigm in Cognitive Science, and others have attempted to follow up on this [3]. Now, in any such attempt, the notions that cognition is embodied, embedded (it is more usual to say "situated"), and extended undeniably play key roles. So as a proponent of "existential enaction," I cannot protest against the association of enaction with the other three "Es." However, what I can and do protest about is adding in "enacted" as an ancillary element at the end of the list. In my view, these three E's are *subservient* to the overriding theme of enaction. Mixing them up indiscriminately, in the way that is done by proponents of the "4Es," leads to missing the wood for the trees.

To sum up, then, I would like to conclude by an invitation. As I have tried to explain, the neuroscience of consciousness potentially opens up a breach in our normal functioning as scientists; it offers the opportunity for those so inclined to introduce a subjective, existential dimension into their work. Of course, one can bring a horse to the water, but one cannot force him to drink. This is particularly so here, where intimate personal attitudes are at stake. But I hope that I have done enough to make the invitation appealing.

Author details

John Stewart
Centre de Recherche Enaction Design (CRED), Technological University of Compiègne, Compiègne, France

*Address all correspondence to: js4a271@gmail.com

References

[1] Chalmers D. Facing up to the problem of consciousness. Journal of Consciousness Studies. 1995;**2**:200-219

[2] Varela F. Neurophenomenology: A methodological remedy for the hard problem. Journal of Consciousness Studies. 1996;**3**:330-335

[3] Stewart J, Gapenne O, Di Paolo E, editors. Enaction: Toward a New Paradigm for Cognitive Science. Boston: MIT Press; 2010

[4] von Uexküll J. A stroll through the worlds of animals and men: A picture book of invisible worlds. Semiotica. 1992;**89**(4):319-391

[5] Husserl E. Ideas Pertaining to a Pure Phenomenology and to a Phenomenological Philosophy—First Book: General Introduction to a Pure Phenomenology. The Hague: Nijhoff; 1913 Translated by F. Kersten, 1982

[6] Petitmengin C. Describing one's subjective experience in the second person. An interview method for the science of consciousness. Phenomenology and Cognitive Sciences. 2006;**5**:229-269

[7] Jaynes J. The Origin of Consciousness in the Breakdown of the Bicameral Mind. Boston: Houghton Mifflin; 1976

[8] Menary R. Introduction to the Special Issue on 4E Cognition. Phenomenology and the Cognitive Sciences. 2010;**9**:459-463

Chapter 3

Influence of Gut Microbiota on Behavior and Its Disturbances

Valentina Ignatova

Abstract

Hippocrates statement that "All disease begins in the gut" continues to be up to date more than 2000 years later. Growing number of scientific reports focus on the important role of intestinal microorganisms for modulation of many systems and human behavior. As a key component of the gut brain, gut microbiota influences the development and maturation of the hypothalamic-pituitary-adrenal axis, affects the development and function of the immune system, regulates the blood-brain barrier, modulates the synthesis and recognition of neurotransmitters, regulates neurogenesis, formation of myelination and supports the development and function of the brain. Disruption of gut-brain axis function is associated with alterations in the stress response and might contribute to neuropsychiatric diseases as depression, autistic spectrum disorders, rapid eye movement sleep behavior disorder, Parkinson disease, Alzheimer disease and other mental conditions. Studies in animal models are crucial for guiding research on brain-gut-microbiome axis in humans, as the impact of microbiota on specific brain regions and aspects of animal behavior will help in the selection of tasks for cognitive assessment. Exploring the interaction of gut microbes and human brain will not only allow us to better understand the pathogenesis of neuropsychiatric disorders, but will also provide us new opportunities for the design of novel immuno- or microbe-based therapies.

Keywords: gut microbiota, brain, gut-brain axis, behavior disturbances, modulation, human behavior, animal models

1. Introduction

Hippocrates statement that "All disease begins in the gut" continues to be up to date more than 2000 years later. The fields of microbiology, gastroenterology and neuroscience have evolved gradually over time and remarkable progress in modern medicine has been achieved not only in their individual trajectories, but also in their active interaction. It has recently become evident that gut bacterial flora can greatly influence all aspects of physiology, including gut-brain communication, brain function and even behavior [1].

The population of microorganisms, localized in the human gut and consisting of bacteria, viruses, protozoa, fungi etc., definitely exceeds the number of cells that make up the human body. The collection of these microorganisms, their genomes and the factors that they produce are all part of the gut microbiome [2, 3]. The role of microorganisms that make up the intestinal flora can be identified as pathogenic, neutral, or useful for the host. The beneficial bacteria known as probiotic bacteria predominate in the intestine of healthy subjects. The word probiotic has

Greek origin and its meaning states "for life". In fact, probiotics are referred to live microbes which are important for maintaining the intestinal microbial balance and have the capacity to keep and improve the health of their human host" [4–6].

The intestinal microbiota and its metabolites influence modulation of gastrointestinal (GI) functions through their ability to affect gut permeability, mucosal immune function, intestinal motility and sensitivity, and also activity of the enteric nervous system (ENS) [6, 7]. Multiple mechanisms, including endocrine and neurocrine pathways, are suggested to be involved in gut microbiota-brain signaling. On the other hand, the brain can in turn alter microbial composition and behavior via the autonomic nervous system (ANS) [8].

Evidences from studies in rodents raised in a germ-free (GF) setting pointed that the gut microbiota influences the development of emotional behavior, stress- and pain-modulation systems and brain neurotransmitter systems. Furthermore, perturbations of microbiota occurring as a result of probiotics and antibiotics application exert lead to effects on some of these modalities in adult animals [8–11].

The absence of micro-organisms in the gastrointestinal tract (GIT) of mice shows a reduction in the number of Peyer's patches and IgA producing B-cells in the lamina propria versus healthy controls, whereas the introduction of microbes reverses these effects [12, 13]. It is curious that, GF mice also provide evidence of an overactive hypothalamic-pituitary-adrenal (HPA) axis and reduced monoaminergic activity, suggesting that microbial colonization can have lasting effects on central systems, which are involved in the psychopathology of depression [14].

Some of the most common species/probiotics are bifidobacteria. Shortly after birth, up to 90% of the bacteria found in children's GIT are bifidobacteria, and in adults they still account for approximately 3–5% of the microflora [15]. Moreover, in inflammatory diseases such as irritable bowel syndrome (IBS), treatment with bifidobacteria normalizes the existing disequilibrium between pro-inflammatory and anti-inflammatory cytokines in this disease [16, 17]. Based on the established important role of the balance between anti- and pro-inflammatory cytokines in the pathophysiology of depression [18, 19], it can be hypothesized that probiotics may have potential antidepressant properties. Of course, the potential benefits of probiotics as adjuvant therapy in depression are currently being discussed [20]. A recent study of Benton et al. has demonstrated a beneficial effect of long-term probiotic treatment on the mood of healthy subjects [5, 21].

It is supposed, that the violation of the two-way functional connection between brain and gut microbiota take a part in the pathogenesis of certain diseases of "gut-brain-axis" such as IBS and impairments of GI-functionality [1, 22] but it could be also involved in the pathogenesis of a lot of significant neuropsychiatric diseases: autism spectrum disorders (ASDs) [1, 23], Parkinson's disease [24], mood disorders [25]; and chronic pain conditions [1, 5, 26].

Unfortunately, the information how these findings could be transferred to healthy humans or to disease states involving the brain or the gut-brain axis is still insufficient. Further research with focus on this topic for translation to human physiology and to diseases such as irritable bowel syndrome, autism, anxiety, depression, and Parkinson's disease should be performed [8].

The interaction between gut microbiota and brain at the levels of gut-brain axis and their influence on manifestation of gastrointestinal, mental and neuropsychiatric diseases is presented at **Figure 1**.

The aim of the review is to examine the dependence between the functioning of microbiota-gut-brain axis and human behavior and how it can contribute to a better understanding of human psychology and choosing an appropriate therapeutic approach in cases of behavior disturbances.

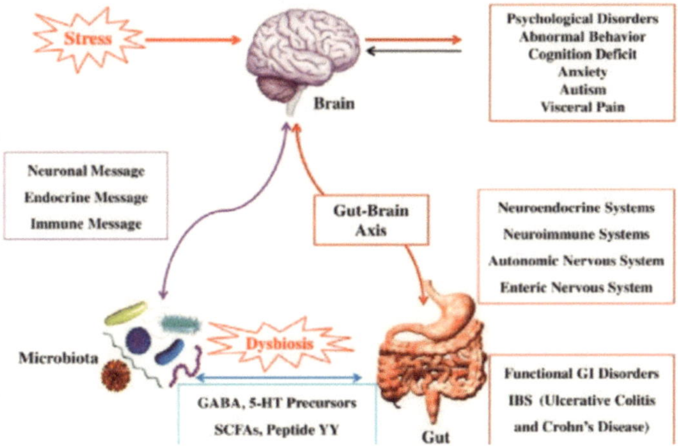

Figure 1.
Bidirectional interactions between gut microbiota and brain. Relationship with medical conditions [27].

2. Method of searching

An advanced search was performed in electronic database (PubMed, MEDLINE), based on the combinations of the following key words and phrases as entry screening criteria: "gut microbiota", "brain", "behavior disturbances", "modulation", "microbiota-gut-brain axis", "influence of gut microbiota on human behavior", "abruption of microbiota-gut-brain axis", "animal models of interrupted microbiota-gut-brain axis". The time period of the search was unlimited. The relevant information was selected from systematic reviews, meta-analysis, books chapters, original articles, conference abstracts and theses, published in English. The exclusion criteria targeted case reports which were not entered in the analysis. The reference of the related scientific sources from the selected articles was additionally checked and the relevant of them were also included in the recent work. A relationship between functioning of intestinal microbiota and human behavior was searched. Based on the above listed criteria, initially were retrieved 3 books,

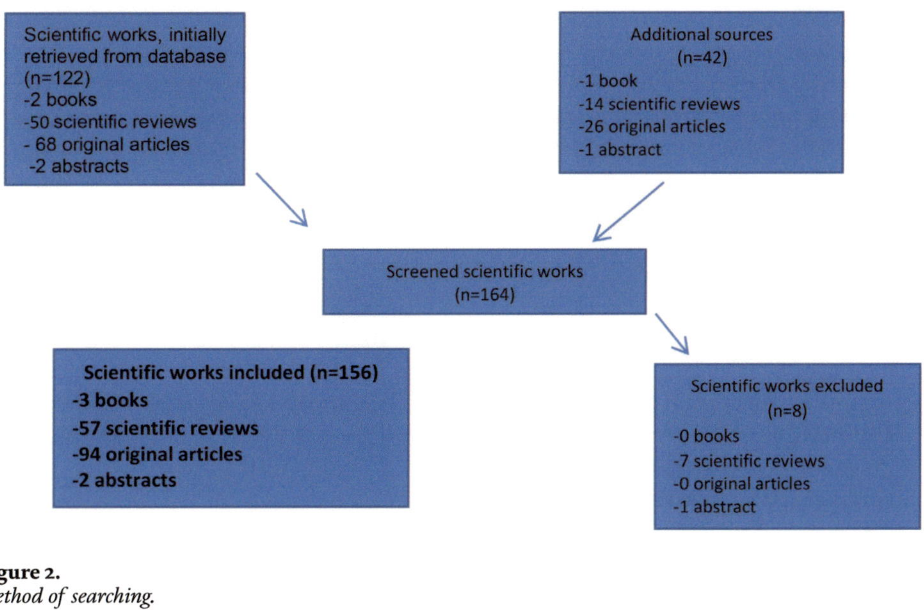

Figure 2.
Method of searching.

60 scientific reviews, 98 original articles, 3 abstracts. The final number of selected and really used scientific works was as follows: 3 book chapter, 57 scientific reviews, 94 research studies, 2 paper abstracts on the current topic "Influence of gut microbiota on behavior and its disturbances"—**Figure 2**. The chronology of knowledge regarding microbiota-gut-brain axis and its influence on human behavior was also tracked. The similarities of points of view and results, as well as the contradictions in published literature were analyzed. The possibilities for psychological and drug interventions in patients with behavior disturbances based on the summarized conclusions were also highlighted.

3. Role of gut-brain axis for mental health

The essential interaction between the gut and the brain through the gut-brain axis is well established. The environment and related factors render influence on central nervous system (CNS), as well as on HPA axis. Furthermore, the CNS interacting with the ENS, the intestinal muscular and mucosal layers via vegetative afferent and efferent tracts, modulates gut functions as permeability, mucus secretion, motility, as well as host immunity [13, 28, 29]. Thus, CNS inputs can affect the gut functions, while gut inputs could modify specific CNS processes [1, 30]. Interruption of these bidirectional interactions may provoke neuroinflammation processes and could be involved in the pathogenic ways responsible for development of CNS disorders [13].

3.1 Anatomy of the gut-brain axis

The functioning of the colon is modulated by both internal (intrinsic) and external (extrinsic) neural pathways [31].

3.1.1 Intrinsic innervation

The ENS is integrative system of neurons with structural complexity and functional heterogeneity, similar to these of brain and spinal cord. Its main role is to control motility, secretion, mucosal transport and blood flow of the GIT [32]. The ENS realizes these functions via motor neurons, localized in ganglia, composing a final common pathway to the effector cells of the GIT. Although these specialized motor neurons receive some impulses from CNS through parasympathetic and sympathetic pathways, their function is predominantly coordinated by sensory neurons and interneurons localized within the ENS [30].

3.1.2 Extrinsic innervation

3.1.2.1 Splanchnic "sympathetic" nerves

Their noradrenergic fibers within the wall of the GI tract arise from cell bodies embedded in the prevertebral sympathetic ganglia. The major "sympathetic" projections to the large intestine originate from the inferior mesenterial ganglia and the remaining noradrenergic fibers to the rectum are provided by the pelvic ganglia.

3.1.2.2 Vagal (parasympathetic) innervation

The vagal nerve transfers information between the internal organs and the brainstem. It contains both afferent and efferent nerve fibers and innervates the entire gut with exception of the distal third of the colon. Vagal afferents terminate in the nucleus of the solitary tract (NTS) whose impulses go up through the parabrachial nucleus (PBN) to the thalamus, limbic system and insula [33, 34]. Spinal fibers pass up via the spinothalamic tract and the dorsal spine columns. Respectively, the spinothalamic pathway goes ascendingly to the thalamus, and the dorsal columns give projections to the gracile nucleus and cuneate nucleus in the upper medulla. Efferent impulses from the last rostral medullar structures reach the thalamus through the medial lemniscus. In turn, thalamic projections ascend to the primary somatosensory cortex and insula [30]. Vagus motor nuclei are represented by nucleus ambiguus (NA) and dorsal motor nucleus (DMN). The DMN is a source of efferents to the smooth muscles of the gut that form synapses with the neurons of the MP [7, 33].

3.2 Gut microbiota influences and human brain function

3.2.1 ANS modulation of the gut microbial environment

Impaired intestinal transit caused by compromised migratory motor complexes (under parasympathetic modulation) is associated with an increased microbial colonization in the small intestine [5, 35–37]. The frequency of regular migrating motor complex is influenced by the number of feeds, quality of sleep and stress. Acute stress is associated with increased parasympathetic output to the small and large intestine and decreased vagal output to the stomach [38]. ANS affects the mechanisms of immune activation at the level of intestinal epithelium. This process could be influenced trough modulation of the intestinal immune cells such as macrophages and mast cells against gut luminal microbes through antimicrobial peptides or indirectly via changing the access of gut microbiota to the intestinal immune cells [1, 39]. This process could be influenced trough modulation of the intestinal immune cells such as macrophages and mast cells against gut luminal microbes through antimicrobial peptides or indirectly via changing the access of gut microbiota to the intestinal immune cells.

Preclinical studies proved that increased permeability of intestinal epithelium after exposure to different stressors is result of easier translocation of gut microbiota followed by driving of immune response [1, 40, 41].

3.2.2 Effects of host's signal molecules on the function of the intestinal microbiota

Neuronal and neuroendocrine signaling molecules as catecholamines, serotonin, cytokines, GABA, dinorphine, etc. dispersed into the gut lumen through neurons, immune and enterochromaffin cells can also play a role in the modification of intestinal environment [1, 8, 42], probably regulated by the CNS [43]. Many stress factors result to increasing of both plasma and luminal gut levels of catecholamines as norepinephrine [8, 44]. In vitro experiments indicated that some pathogen microbes could change their proliferative ability after exposition to exogenous catecholamines. Norepinephrine may accelerate the proliferation of several strains of intestinal pathogenic microorganisms and could increase the virulence of *Campylobacter jejuni* [1, 45, 46].

3.2.3 Microbe-to-host signaling by microbial signaling molecules

Metabolites produced by intestinal microbiota, including short chain fatty acids (SCFAs), bile acid metabolites and neuroactive agents such as GABA, tryptophan precursors and metabolites, serotonin and catecholamines, including free metabolites and cytokines released during the immune response to microbes may deliver the signal to the host via local cell receptors in the intestine [37, 47]. These factors can also give signals via neurokines through afferent vagal and spinal pathways and endocrine mechanisms to target non-GI tracts, including vagal afferents in the portal vein and receptors in the brain. A significant part of the metabolites identified in the circulation are with intestinal microbial origin [48, 49]. The enteroendocrine cells, as well as the neurons forming the submucosal and myenteric ganglia, express different types of SCFA receptors [50] A diet that includes *Bifidobacterium breve* leads to elevated levels of fatty acids in the brain, but unfortunately there is not a clear explanation for that mechanism [1].

Actual studies confirm that multiple nuclear receptors (NRs) are expressed in the GI tract and several microbe-produced metabolites act as ligands of NRs. Intestinal bacteria secrete metabolites including indole derivatives, hormones and secondary BAs, which play role of natural ligands for the host's NRs [51]. In this way the microbial metabolites can realize biological effects in the human body via regulation of the host's gene expression. These dynamic interactions permit overall control on the health or disease development in the host through direct effect of the microbiota on the human physiology [11]. Via signaling to the brain, the microbiota regulates metabolism, CNS development, inflammation as well as mood and behavior. It is essential that the human host has the ability to voluntarily influence and meliorate its own microbiota through nutritional or probiotic interventions [52].

Latest research found that, in the presence of the microbiota, the intestinal epithelial lining generates physiological levels of oxidative stress. On the other hand, these interfere both with the composition and functionality of the microbiota (e.g., anaerobes thrive in the presence of electron acceptors) and directly with the gut permeability. This lead to increased probability of xenobiotic molecules to reach the systemic circulation as well as the CNS [53].

Another well-known interaction between the microbiota and CNS involves astrocytes. Astrocytes represent a functionally diverse group of glial cells, which are responsible for ion homeostasis, neurotransmitter balance, storage of glycogen, the integrity of the blood brain barrier (BBB), realizing of the neuronal signaling, which play a main role in the neuroinflammation process [54]. The inflammation could be suppressed by modulation of type I interferon signaling in astrocytes, initiated by microbial metabolite products, which activate the aryl hydrocarbon receptors (AhRs), The indoles, released from gut microbiota, act as AhR agonists [55]. The dietary tryptophan in intestinal cavity, which is undigested, is transformed into indole in the presence of the microbial enzyme tryptophanase. Then the indoles could be modified through microbial or viler enzymes to indole derivate with various affinity [10, 56].

4. Dysregulation of the gut-brain axis. Evidences from experiments on animals

Alterations in the microbial contain of the GIT are considered to contribute to inflammatory and functional bowel disorders and psychiatric comorbidities. The results of recent studies using various strains of mice and rats, various strains of probiotics and various experimental paradigms reported a number of microbial bowel modulation effects in emotional behavior [57–59], learning and memory [60], social interactions [61] and nutritional behavior [62].

4.1 Experiments on mood changes in disturbed gut-microbiota in rats

The first experiments in young GF mice which confirmed the influences of gut microbiota on postnatal development of the hypothalamic-pituitary response to stress were performed by Sudo et al. in 2004. It is interesting, that the GF mice expressed reduced anxiety-like behavior in the elevated plus maze (EPM), a reliable behavioral test that examines approach and avoidance behavior in mice, compared to specific pathogen free (SPF) mice [63].

Desbonnet et al. evaluated the potential antidepressant properties of probiotics through testing of rats chronically treated with Bifidobacteria infantis in the forced swim test [5]. Probiotic administration in naive rats had no effect on swimming behaviors. However mitogen stimulation in probiotic treated rats lead to substantial reduction of IFN-γ, TNF-α and IL-6 cytokines compared to controls ($p < 0.05$). In addition, the plasma concentrations of tryptophan ($p < 0.005$) and kynurenic acid ($p < 0.05$) were significantly elevated in the rats, treated with bifidobacteria [20]. Treatments with Bifidobacteria also lead to reduced 5-HIAA concentration in the frontal cortex and a decrease in DOPAC in the amygdaloid cortex [5].

Schroeder et al. provided evidences for production of benzodiazepine ligands in a rat model of encephalopathy or butyrate acting as a histone D-acetylesterase that was shown to have an antidepressant effect [64, 65].

The study of Arseneault-Breard et al. gave the first evidences for beneficial effect of probiotics *L. helveticus* R0052 and *B. longum* R0175 on post-myocardial infarct depression in rats. This positive probiotic influence was engaged in maintaining of the gut barrier integrity, which is possibly associated with the host' inflammatory state after MI [84].

The association of increased HPA axis responses and reduced anxiety-like behaviors observed in several of the studies performed in GF mice suggests that HPA axis and nonhypothalamic (anxiety-like behavior) components of central stress circuits may be affected on different ways according to the GF conditions, depending on species and mouse strain. These findings suggest that the increased HPA axis activity in GF animals may represent a response of the organism to the loss of microbiota-related energy sources [8].

Savignac et al. demonstrated that the two *Bifidobacterium* strains used in their study were able to improve the anxious phenotype of innately anxious BALB/c mice in a strain-specific manner and the effect was better than that from the administered antidepressant escitalopram. These findings support the statement that probiotics could be a reliable alternative for treatment of mood disorders [142].

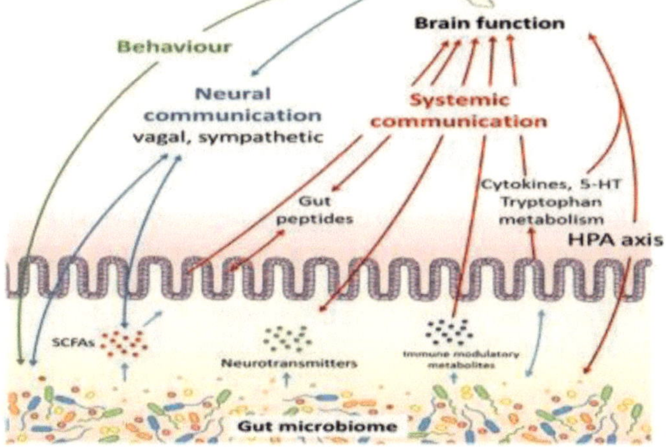

Figure 3.
Impact of gut microbiota and external stress factors on behavior [66].

On **Figure 3** is presented influence of both gut microbiota and external stressors on behavior.

4.2 Experiments on behavior changes in disturbed gut-microbiota in rats

The increased hippocampal brain-derived neurotropic factor (BDNF) registered in the ATM-treated mice is corresponding with their gregarious behavior. A recent study found increased BDNF expression in the amygdala during fear learning [67, 68]. Over activation of the amygdala also has been implicated in depression and anxiety [67, 69]. Lower levels of BDNF in the amygdala of ATM treated mice were associated with increased exploratory behavior.

Bercik et al. found that SPF mice who received antimicrobial agents per os demonstrated enhanced exploratory behavior and hippocampal expression of BDNF. This finding was associated with temporary alteration of the representatives of their microbiota and was not accompanied by inflammatory status, alteration of gastrointestinal neurotransmitters levels, nor with vagal or sympathetic function. Intraperitoneal application of antimicrobial agents to SPF mice, similar to their oral administration in GF mice had no influence on behavior. Increased exploratory behavior and high hippocampal levels of BDNF were reported in GF BALB/c mice, colonized with microbiota from NIH Swiss mice. Suppression of exploratory behavior was demonstrated in GF NIH Swiss mice, colonized with BALB/c microbiota [2, 70].

The study of Bercik et al. did not provide proof for intestinal inflammation, as oppose to Verdú et al.' investigation [71], where administration of ATMs in a higher dose and for a longer period was made in NIH Swiss mice. In the Bercik's experiment embarrassment of the intestinal microbes did not change myeloperoxidase activity, histology or cytokine profile of the colon [8]. No differences in serotonin, dopamine, or noradrenaline content in the gut of ATM-al. treated mice were observed, suggesting that these neurotransmitters are not involved in mediating the behavioral changes observed in the model.

Li et al. and Bercik et al. reached similar results on memory and learning skills in adult mice [11], applying different nutritional supplements to animals at a very early age with the following disruption of the intestinal flora in very young age. Working and referred memory was better in the animals on rich in beef diet as opposed to the mice on standard meal [8].

Neufeld et al. supposed that the low anxiety-like phenotype was accompanied by long-term changes in plasticity-related genes in the hippocampus and amygdala. They found altered GF behavior, accompanied by a decrease in the N-methyl-D-aspartate receptor subunit NR2B mRNA expression in the central amygdala, increased BDNF expression and decreased serotonin receptor 1A (5HT1A) expression in the dentate granule layer of the hippocampus. It is the first work which demonstrated an altered behavioral phenotype related with lack of gut microbiota [59].

In their work Bravo et al. registered increased levels of GABAB1b mRNA in cingular and prelimbic areas in mice treated for a long time with *L. rhamnosus* (JB-1), while the concentration of these neurotransmitters was reduced in the hippocampus, amygdala and locus coeruleus in the same experimental animals. Furthermore, the GABAAα2 level was reduced in the prefrontal cortex and amygdala, and increased in the hippocampus. The observed mice expressed reduced response to stress, associated with releasing of corticosterone. Similar neurochemical and behavioral effects were not expressed in mice, who has underwent vagotomy [12, 73].

In their study Park et al. demonstrated that depressed-like behavior in mice that underwent bilateral olfactory bulbectomy (OBx) was associated with altered colonic motility and a shift in the microbiota profile. Their experiment also supposed that increased prokinetic neuropeptide, gut hormone and serotonin in the colonic wall are mediators of the altered motility [25]. Their finding was consistent with those of Rodes et al. who showed changed colonic transit and altered stability of the colonial microbial community [74].

Hsiao et al. demonstrated GI barrier defects and microbiota alterations in the maternal immune activation (MIA) mouse model who displayed ASD signs. MIA generation, who has received *Bacteroides fragilis* (human commensal microbe) per os, has evolved altered bacterial gut content which predisposes to impaired communication and manifestation of stereotypic, anxiety and sensorimotor behavior. The described experimental model showed change in profile of the serum metabolites and their levels. It is other evidence for the gut microbiota impact on human behavior through the gut microbiome-brain functional axis and it could help in searching of relevant probiotic treatment of behavior disturbances in neurodevelopment diseases in human [3, 75].

It was found that gut microbiota status reduce social interactions in GF mice and probiotics improve social interactions in a post-MI rat model. Desbonnet et al. examined whether the degree of information transfer during social interaction is disrupted in GF mice. In their experiment GF mice spent a decreased proportion of time engaged in social investigation and substantially greater proportion of time engaged in repetitive self-grooming behavior during social interaction. After GF bacterial colonization these behaviors were normalized, which is evidence for involvement of microbiota in modulation of such behaviors. However, the quality of information transfer during the interaction was not affected in GF mice, indicating that the ability to process information per se during social interaction was not affected in GF mice [76].

It is important to note that many of the psychologic deficits, registered in GF mice, are specific to males in which higher incidence of neurodevelopmental disorders was registered compared to females [59, 63, 77–79]. de Theije et al. demonstrated that gender-specific inflammatory conditions are present in the small intestines of VPA in utero-exposed mice and are accompanied by a disturbed serotonergic system both in the brain and in the intestinal tract [80]. Gut microbiota-associated behavioral changes were reported in different ASD mouse models using valproic acid administration or maternal infection; in the latter instance some behavioral disorders were favorably influenced by probiotic therapy with *Bacteroides fragilis* [8, 9].

Several studies proposed the influence of intestinal microorganisms on eating behavior [80], probably as a consequence of modified fatty acid receptors, gut receptors, responsible for taste, alteration of the intestinal transportation mechanisms or disturbed releasing of satiety hormones [9, 81, 82].

Crumeyrolle-Arias et al. found that lack of intestinal microbiota in sensitive to stress strain rats lead to neuroendocrine and behavior reactions of acute stress and significant changed degree of the dopaminergic turnover in the higher brain structures which modulate stress and anxiety—another support for the crucial impact of the gut microbiota on the higher brain activities [9, 15, 83, 84].

Recently it was reported that impaired permeability of the blood brain barrier in GF mice probably will restrict reaching of the liver bacterial metabolites to the brain [85]. Numerous remodeling experiments in GF animals confirmed that deviations of brain metabolism and behavior could be preserved through reconstitution of the gut microbial composition [1, 86].

Wong et al. found that genetically determined caspase-1 deficit in mice suppresses the anxiety-depressive like behavior and improves the motor activity and locomotor abilities, as well prevents manifestation of depressive symptoms after chronic exposition at stressors. On the other hand, minocycline as pharmacological antagonist of caspase-1 alleviates the depressive like symptoms in wild type mice provoked by stress. Actually, both chronic stress and pharmacological inhibition of caspase-1 modify the composition of fecal microorganisms almost in the same way [3, 87].

The GF model has some limitations, which suppose that the investigators should be cautious in extrapolating the conclusions obtained in animals on people. Important is the fact that GF mice are born under aseptic conditions, such as separation from the mother via cesarean section and directly placement of the newborn in an special insulator in which the air, in which everything is sterilized, including the air, food and water. The biochemistry of brain and gut intestine is quite different [1, 81], HPA axis responses [63], in emotional, [58], social [75, 79], metabolic function, and ingestive behaviors [82] between GF animals and control animals which contain normal or pathogen-free flora obtained by colonization from the mother [8, 78]. However, up to date studies with animal model proved that the gut microbiota can influence the central nervous system in the absence of substantial changes in local or circulating cytokines or specific intestinal neurotransmitter.

It is unambiguous that bacterial products can get access to the brain via the bloodstream, they can act through the immune system via cytokine releasing by the mucosal immune cells, or through the endocrine system by triggering gut hormone release from enteroendocrine cells [9, 87]. Since GF animal models are not analogous to the development of the human brain, premature conclusions about the significance of these findings to humans should be avoided [88].

5. Findings from clinical, imaging and neurophysiological tests on the human brain-gut axis

Gut to brain pathways have been explored through cortical evoked potentials (CEPs), magnetoencephalography (MEG), positron emission tomography (PET), and functional magnetic resonance imaging (fMRI) [30].

Loening-Baucke et al. applied anorectal CEP in children with constipation and encopresis and found significantly prolonged latencies of the early-onset potentials, suggesting a defect in afferent pathway conduction [89].

The perception of painful stimuli is accompanied by activation of anterior cingulate area in people in a healthy condition, as opposite to subjects suffering from IBS in which activation of left prefrontal cortex occurs probably due aberrant CNS processing [5] Subsequent research in patients with IBS also suggest that rectal hypersensitivity induced by repetitive distention of the sigmoid colon correlates significantly with increased blood flow in the thalamus and that an aberrant thalamic response to pain could be the reason for the abnormal sensitization.

In studies of Ertekin et al. and Herdman et al. conducted at different times reproducible EMG responses on the part of external anal sphincter were evoked by cortical magnetoelectric stimulation. Turnbull et al. managed to differentiate the topographic areas of the external anal sphincter and the pelvic floor muscles represented at the medial side of the primary motor brain cortex using TCMS. This representation is bilateral and shows asymmetry in some individuals [7, 30, 90–92].

6. Effect of interventions targeting the gut microbiota

Known approach for registration the effects of intestinal microbes on brain function is the use self-reporting measures to determine how the brain function alters under the influence of induced from probiotics microbial proliferation.

The level of anxiety and psycho-emotional stress was reduced in human (both male and female), treated with *Lactobacillus* and *Bifidobacterium* versus persons who toke control substance in a randomized placebo-control trial. However other study using different strain *Lactobacillus* does not succeed to confirm this conclusion.

But another study using a different *Lactobacillus* probiotic, failed to confirm these findings [21, 93]. The limitations in the study design including the size of the cohort, the mood of the surveyed contingent, the used assessment tools, the inter-individual differences in microbial composition and the differences between the probiotics may be the cause of the discrepancy in the results.

Another approach is to use functional MRI (fMRI) to assess changes in the human brain in response to gut microbial modulation. One study showed that chronic ingestion of a probiotic consortium altered functional brain responses in healthy women [94]. In this study, the answer to the emotional face recognition task was measured with fMRI in healthy women before and after intake of active probiotic for 4 weeks, unfermented dairy product or no treatment. Women who were treated with probiotics demonstrated diminished response to the task of emotionally recognizing in extensive brain networks, including territories, responsible for sensation and emotions. Self-assessment of anxiety and depression was not significantly different between in between the studied groups. But altered fMRI responses proposed a substantial change in response to emotional negative stimuli. Separate functional brain imaging study explored the modulatory impact of gut microbiota in subjects with mild cognitive impairment and hepatic encephalopathy through administration of non-absorbable antibacterial agent [95]. More successful coping with the cognitive task corresponded with increased subcortical activity and better frontoparietal connectivity on fMRI. Other investigation with antibiotic administration in people with the same diseases during 8 weeks also confirmed improved cognitive level and established altered serum metabolites with supposed bacterial origin [1, 26, 96].

7. Role of microbiota-gut-brain axis in neurological and psychiatric diseases

Changes in the microbial environment, as a result of different stressors, are linked with alterations of barrier, motility and activation of the immune system. Perturbation of this axis lead to changes in the stress-response and behavior, which are thought to be involved in several CNS diseases, such as anxiety, depression autism, Parkinson's disease and Alzheimer's disease [97].

Neuropsychiatric comorbidity, including depression and anxiety, is common finding in patients with a functional GI disorder such as the IBS and it reaches 60% of this somatic pathology. On the other hand, IBS has also been related with changes in the gut microbiota including reduced diversity and temporal instability at the genus level. It is interesting to note that behavioral and psychological changes are often present in patients with active celiac disease, which are associated with findings of regional cerebral hypoperfusion in their brains [25, 98].

Some recent works reported for changed expression of GABA A receptor and its B subunits, responsible for the primary inhibitory brain mediation [73, 99, 100], subunits of NMDA receptors, realizing excitatory neurotransmission [101], concentration of serotonin 1A and tryptophan. Some of the above mentioned alterations corresponded with disturbed emotional behavior, which supports the interaction between microbial composition and behavior [1, 8].

In recent years evidence has emerged that neurodegenerative diseases (NDs) are strongly associated with the microbiome composition in the gut [101, 102].

7.1 Role of microbiota-gut-brain axis in mood disorder

A "leaky gut" is suggested to play a pathogenic role in depression. There are evidences for altered intestinal permeability in patients with mood disorder and their first degree relatives [18]. For most of the depressed patients the brain-gut axis function is impaired, including imbalance in brain neurotransmitters, decline of brain neuroplasticity, dysfunction of hypothalamic-pituitary-adrenal axis, chronic periphery inflammation and neuroinflammation, as well as gastrointestinal diseases and gut microbiota dysbiosis [103]. However, the impact of depression on the microbiota has not yet been studied [25].

Wong et al. proposed that suppression of caspase-1 plays a protective role in modulation of the interaction between representatives of the intestinal microbiota and the state of stress. They reported the importance of signals from inflammasome along the gut microbiota-inflammasome-brain axis which attribute to modification of cerebral processes, especially for manifestation of anxiety-depressive symptoms [3, 87].

Acute tryptophan depletion (ATD) in subjects suffering from depression, was preceded by bimodal emotional processing, corresponding with bimodal manifestation of the clinical symptom. It was proved in a small patient's cohort where the alleviation of depressive symptoms occurred 24 h after ATD and the mood processing was at the better level about 5 h after depletion. The opposite processes were registered in patients who experienced worsening of the depressive symptoms [2, 104].

Serotonin is a key element of this axis, acting as a neurotransmitter in the CNS and in the ENS, located in the gut wall. This transmitter is formed in neuroendocrine cells and plays a role of paracrine hormone in the intestine. Serotonin is also an endocrine hormone which passes into the blood and bind to the platelets. Besides its system effects like maintaining the bone density and participation in metabolite processes, it performs a key connection between both ends of the gut-brain axis [105]. It is interesting that most of the serotonin is produced at the periphery predominantly in the tGI epithelium, but also in bones, breast and pancreas. Only 5% of its synthesis is realized at central level. The only difference in ways of serotonin synthesis is the use of tryptophan hydroxylase type 1 in peripheral mechanism and instead of it-type 2-in the central one [105]. The reversal process of serotonin degradation is performed with the help of monoamine oxidase and aldehyde dehydrogenase to 5HIAA both in the periphery and in the CNS [2, 96, 106].

7.1.1 Microbiota-gut-brain axis in major depression

Major depression disorder (MDD) is an incapacitating multifactorial psychiatric disease, which is characterized by a range of symptoms affecting both emotional and cognitive domains [107]. The hypothesis of activated peripheral blood monocytes and T lymphocytes is well known [18]. Another supposed mechanism is impaired excitation/inhibition balance that is potentially mediated by the reduced amount of GABA. The low concentration of brain-derived neurotrophic factor has been proposed as a unifying hypothesis for reduced cell numbers in frontal cortex and amygdala and also for reduced hippocampal volume. Despite the great advances in the knowledge of this disease, its etiology and pathophysiology are still not fully understood [108].

It is important that metabolites as hippurate, dimethylamine and dimethylglycine derived from the blood of patients with MDD are actually products of their intestinal microbiota [109]. Similar to findings in animal experiments in depression, limited number of studies in humans with small cohorts found changes of the gut microbiota strains [25, 110]. All this trials proposed the potential relationship between the alteration of gut microbiota composition and MDD manifestation [18].

Significant difference in the isolated bacteria from stool samples of 58 Chinese subjects, diagnosed with major depression compared to 63 healthy individuals was found. Y.

The following three main bacterial phyla are specific for the gut microbiota of depressed subjects: Firmicutes, Actinobacteria and Bacteroidetes. Depression behavior models were created through transplantation of stool samples taken from five subjects with depression into germfree mice. And vice versa, transplantation of feces from five healthy individuals did not lead to behavioral effect. Mice receiving microbiota from patients with depression showed disturbances in hippocampal gene activation and also in carbohydrate and amino-acid metabolism [16, 109]. This study provided convincing evidence that the depressive phenotype could be transferred by transplantation of the microbiota.

Kelly et al. recruited 34 patients with major depression and 33 healthy individuals with similar demographics. Plasma levels of cytokines, C-reactive protein, salivary cortisol and plasma lipopolysaccharide-binding protein were determined by ELISA, and showed alterations supporting a proinflammatory phenotype linked with depression. Depression was associated with decreased gut-microbiota richness and diversity. A fecal microbiota transplant was prepared from a subgroup of patients with depression or from healthy individuals and transferred by oral gavage to a microbiota-deficient rat model [114].

But further research in larger samples and unified MDD populations is required to confirm whether disturbances in gut microbiota have a causative role for the onset of MDD.

7.2 Microbiota-gut-brain axis in autistic spectrum disorders (ASD)

During the early onset of this developmental disorder an autistic enterocolitis and changes in intestinal permeability occur [111]. Moreover, urinary metabolic phenotyping has determined biochemical changes that were consistent with abnormalities in the composition of the gut microbiota, found in autistic children.

Recent studies suggest that changes in antigenic load due to the impairment of gut barrier function is triggering factor for clinical manifestation of autism [112]. Desbonnet et al. are the first scientists who found that microbiota are crucial for the programming and presentation of distinct normal social behaviors, including social motivation and preference for social novelty, while also being important regulators of repetitive behaviors. Taking into account that these aspects of behavior are impaired in neurodevelopmental disorders such as schizophrenia and autism [5] and with a male preponderance, these data extend our knowledge regarding the genesis of neurodevelopmental disorders of altered sociability. A better understanding of the underlying mechanisms of these social deficits, which may include modulation of immune cell cytokines release, changes in vagal nerve activity and neuroendocrine function, can help for developing of innovative and more effective strategies in managing of these social disturbances [76].

A study of Kang et al. revealed less abundance of Bifidobacteria species and the mucolytic bacterium *Akkermansia muciniphila* in children with autism [114–116]. Other experiment showed less diverse gut microbial composition with lower levels of *Prevotella*, *Coprococcus*, and unclassified *Veillonellaceae* in ASD [117]. Another study showed a significant increase in several mucosa-associated Clostridiales, whereas a decrease in Dorea, Blautia and Sutterella was seen in AUTISM-FGID [118].

7.3 Microbiota-gut-brain axis in Alzheimer disease

Alzheimer disease is a progressive neurodegenerative illness associated with accumulation of proteinaceous misfolded amyloid-b (Ab) fibrils and oligomers, together with neurofibrillary tangles consisting of hyperphosphorylated tau protein in the cerebral cortex and other brain regions [118]. Recent research indicates that alterations of the gut microbiome could activate proinflammatory cytokines and increase intestinal permeability, leading to insulin resistance, which has also been found in AD [119].

Bacterial representatives of the gut microbiome excrete lipopolysaccharides (LPSs) and amyloids. These products lead to forceful pro-inflammatory and innate-immune effects, activate the system Following enhanced amyloid aggregation, as well secondary degeneration occur, which are typical signs of AD, together with impaired cleansing mechanisms of Aβ peptide [17, 120, 121]. It has been suggested that diet and specific nutrients could alert the composition of the intestinal microbiota and might influence the production or aggregation of amyloid proteins [29, 114, 122].

7.4 Microbiota-gut-brain axis in Parkinson disease and its prodromes

Currently it is well established that Parkinson's disease (PD) is not a pure movement disorder of the CNS but also a gastrointestinal disease [115–117], which affects the ENS [123–125]. The main premotor PD symptoms include rapid eye movement sleep behavior disorder, hyposmia, constipation and depression [126].

Rapid eye movement (REM) sleep behavior disorder (RBD) is a parasomnia which results from the loss of physiological motor inhibition and is manifested with abnormal behavior during REM sleep. That disorder leads to varying degrees of complex motor activity which ranges from sleep talking to violent dream enacting behaviors potentially harmful for the subject or bed partner [127, 133].

In their study Heintz-Buschart et al. revealed differential abundances of gut microbial taxa (such as *Akkermansia*) in Parkinson disease (PD) and its prodromal state idiopathic REM sleep behavior disorder compared to healthy controls. The majority (about 80%) of the differential gut strains in patients with PD are very similar to those in subjects with idiopathic REM sleep behavior disorder. Most common are *Anaerotruncus* and *Bacteroides*, which correspond to non-motor symptoms of the disorders. Metagenomic sequencing of specific microbial samples allows the genomic reconstruction [23, 128]. Other studies registered reduction of microorganisms as *Faecalibacterium*, *Coprococcus*, *Blautia*, *Prevotella* and *Prevotellaceae* in gut of subjects, suffering from PD. These alterations are non-disease specific at a lower taxonomic level, for example at phylum stage, but at higher taxonomic level as genus or species, was registered some overlap between alpha synucleinopathies such as PD and multisystem atrophy (MSA) [3, 102].

It has been shown that PD patients with RBD exhibit much higher frequencies of phosphorylated asyn pathology in the colon and in the skin compared to PD patients without RBD [129]. Also, idiopathic RBD subjects exhibit marked pathology in the sympathetic and parasympathetic nervous system, but a relatively intact dopamine system [130].

For *Prevotella* such reduction has also been observed in RBD patients. Based on the attributed functional properties of these bacteria, such alterations could affect gut barrier integrity, short-chain fatty acid (SCFA) production, and inflammation. This would be in line with reports of a leaky gut and reduced levels of SCFAs and lipopolysaccharide binding protein in PD patients.

An interesting link between gut microbiota and asyn pathology could be cross-seeding of amyloid pathology induced by bacterial amyloid proteins such as curli.

So far, human microbiome studies in PD have been carried out exclusively in medicated patients, except for one study that included also idiopathic RBD patients

[131]. While the PD associated microbiome alterations have been confirmed in drug adjusted analyses, confounding effects which could be result of COMT inhibitors intake cannot be excluded. Another potential confounder is colonic dysmotility, which may independently alter microbiota composition [132, 133].

Thus, observed brain and behavioral changes may be mediated by the absence of intestinal microbes directly or indirectly through one or more of the non-brain-related alterations. The latest data show that the intrauterine environment is not sterile, and it can even be supposed that microbial metabolites of the intestine from the maternal microbes of the intestine may have an effect on the development of the fetal brain [75]. The altered signaling of the cecum to the brain secondary to the massive cecal dilation associated with this model may alter the development of the brain regions that process this input [88].

8. Discussion

The gut microbiota has co-evolved with its host for millennia and influences positively many functions of the host organism, as digestion, production of nutrients, detoxification, defense against pathogens and immune regulations [2, 3, 123].

As a key component of the gut brain, gut microbiota influences the development and maturation of the HPA axis [134], affects the development and function of the immune system [135], regulates the blood-brain barrier [136], modulates the synthesis and recognition of neurotransmitters [73], regulates neurogenesis, formation of myelination [137], and supports the development and function of the brain [78]. Microbiota-gut-brain axis plays a crucial role for manifestation of mental disorders [103].

Following the development of gut microbiota, the scientists not only focus on the top-down effects of the brain-gut axis (from brain to gut), but they also devote special attention to the bottom-up influences (from gut to brain) [138]. Alterations of the "gut brain" as pathological changes of intestinal microbiota affect the brain activity and have an impact on behavior. In turn, the emerging brain changes provide feedback on the gut. Uniting the brain and the colon, the brain as targeting gut microbiota organ is becoming a key trend in neuroscience and reliable field for successful management of neuropsychological disorders [11, 103].

Treatment with antidepressants has achieved a significant improvement from the introduction of selective serotonin re-uptake inhibitors and rather the introduced serotonin and noradrenaline re-uptake inhibitors, however, there are still outstanding clinical requirements for the treatment of depression, and better therapeutic strategies are needed, especially with regard to the treatment of depressive cure and additional comorbid painful physical conditions such as GI discomfort [5]. There are confirmations for intestinal microbe changes in patients suffering from major depression [114, 139], as well as in IBS. Considering the serious evidence from laboratory animal models in which the stress affects brain-gut-microbial axis, this area requires more research in humans. In addition to the effects on the immune system, probiotics have also been shown to improve carbohydrate malabsorption [140], which in turn is associated with both early signs of depression and reduced levels of tryptophan [141].

There is a reasonable assumption that probiotic treatment can produce a beneficial effect on mood by raising serotonin precursor levels, tryptophan, and hence increasing serotonin availability [5]. Despite these promising initial findings on microbiota and stress-related disorders, there is a relative lack of research among healthy individuals linking the composition and function of the microbes inhabiting the human intestine and levels of chronic stress or susceptibility to acute stress [26].

Studies in animal models are crucial for guiding research on brain-gut microbiome-axis in humans, as the impact of microbiota on specific brain regions and aspects of animal behavior will help in the selection of tasks for cognitive assessment. Such studies will also be useful in identifying which bacteria may be of particular importance. For example, in rodent models, a specific strain of *Bifidobacterium longum* was found to alter cognition, [142], as well as stress-related behavior and physiology, and a similar effect profile was subsequently observed in people given this strain [26, 143].

Recent research indicates that the gut microbiota is associated with health in the elderly, with those in long-term care having a less diverse microbiota than those living in the community [144]. Even in healthy aging, some aspects of cognition could be deteriorated. Despite the growing interest in this problem, there is still lack of sufficient studies, especially of long-term longitudinal research examining changes in the human gut microbiota with aging. The high inter-individual variation in the gut microbiota also impedes interpretation. Such research efforts should occur in the context of rapid acceleration of genetic sequencing technologies for better characterizing of the gut microbiota [26, 145].

We are witnesses of extraordinary merging of research approaches in different areas of psychiatry, gastroenterology and neurology which significantly improve our understanding of neuropsychiatric diseases and more clearly explain the close relationship between GI and mood disorders. As a result the therapeutic strategies in some mental illnesses significantly advanced. For example, IBS, recognized as linked with psychosocial and GI disorders [6, 146] and often accompanied by depressive symptoms [147, 148] has improved since introduction of an interdisciplinary approach [5]. Although the brain and gut are organs with quite different functions at first glance, the emerging part of the scientific sources provides proofs for their synergy along the "brain and gut axis" and suppose that not only the brain may affect the intestinal function but also the gut, both by direct and by indirect mechanisms can cause alterations in CNS [6, 149], and in stress-related disorders such as depression [5].

It is not known whether the observed changes in the microbiome play a causal role in the development of the intestinal pathology in PD or whether they are a consequence of the altered intestinal function. However, observations that motor symptoms, neuroinflammation, asyne pathology and intestinal motility may be modulated by manipulation of the gut microbes in transgenic asyne-overexpressing mice suggest that such causative effects are possible [150, 151].

In order to establish which mechanisms connect microbe alterations and PD, such studies should use a multiomics approach, including meta-genomic, metatranscriptional and metabolomic assays, in combination with assessment of host factors such as intestinal biopsy, permeability studies, cytokine levels and host genotype. For this purpose multi-center consortia need to cooperate to ensure a sufficient cohort size and standardized methodology [132].

A fuller understanding of the human "hologenome," of human microbial ecosystems and their secretory products should provide a deeper insight into their contribution to age-related neurological diseases associated with amyloidogenesis, CNS inflammation and progressive neurodegeneration [120].

It is suggested that modulating the gut microbiome through specific nutritional interventions and the use of prebiotics and probiotics might represent an effective strategy to reduce the level of chronic inflammation and Ab associated with AD, possibly preventing or ameliorating AD symptoms [29].

A deeper understanding of how psychological development and social and cultural factors affect the brain-gut-microbiome axis will contextualize the role of this axis in humans and give a light on the necessary psychological interventions that will improve the health of the brain-gut-microbiome axis. Interventions apparently aimed at alleviating disorders in a part of the brain-intestinal-microbial axis (e.g., depression psychotherapy) may still affect other parts of the axis (e.g., microbial composition and function) and functional GI disorders such as IBS are disorder of the axis, not an isolated problem of both psychology and gastrointestinal function. Discipline psychology should be aware of these interactions in order to help create a future research program in this emerging research area [26, 152].

9. Conclusion

The gut microbiota influences the brain biochemistry and hence—the behavior irrespective of the autonomic nervous system, specific GI neurotransmitters, or inflammation. The intestinal commensals communicate with the human body via immune, endocrine and neural mechanism. These functional pathways are part of the microbiota- gut-brain-axis and according to preclinical evidence the gut microbes can recruit the above mentioned bidirectional communication relationship to modulate not only the brain development and functioning, but also our behavior. Disruption of gut-brain axis function is associated with alterations in the stress response and might contribute to neuropsychiatric diseases as mood disorder, ASD, REM sleep behavior disorder, Parkinson disease, Alzheimer disease and other mental conditions. Exploring the interaction of gut microbes and human brain will not only allow us to better understand the pathogenesis of neuropsychiatric disorders, but will also provide us new opportunities for the design of novel immuno- or microbe-based therapies.

Conflict of interest

No.

Author details

Valentina Ignatova
Clinic of Neurology, Multiprofile Hospital of Active Treatment, National Heart Hospital, Sofia, Bulgaria

*Address all correspondence to: valyaig@abv.bg

References

[1] Cryan JF, Dinan TG. Mind-altering microorganisms: The impact of the gut microbiota on brain and behaviour. Nature Reviews Neuroscience. 2012;**13**:701-712

[2] Hill CJ, Lynch DB, Murphy K, Ulaszewska M, Jeffery IB, O'Shea CA, et al. Evolution of gut microbiota composition from birth to 24 weeks in the INFANTMET cohort. Microbiome. 2017;**5**(1):4

[3] Sender R, Fuchs S, Milo R. Revised estimates for the number of human and bacteria cells in the body. PLoS Biology. 2016;**14**(8):e1002533

[4] Fuller R. Probiotics in human medicine. Gut. 1991;**32**(4):439-442

[5] Desbonnet L, Garrett L, Clarke G, Bienenstock J, Dinan TG. The probiotic *Bifidobacteria infantis*: An assessment of potential antidepressant properties in the rat. Journal of Psychiatric Research. 2008;**43**:164-174

[6] Matricon J et al. Review article: Associations between immune activation, intestinal permeability and the irritable bowel syndrome. Alimentary Pharmacology & Therapeutics. 2012;**36**(11-12):1009-1031

[7] Forsythe P, Kunze WA. Voices from within: Gut microbes and the CNS. Cellular and Molecular Life Sciences. 2013;**70**(1):55-69

[8] Mayer EA, Tillisch K, Gupta A. Gut/brain axis and the microbiota. The Journal of Clinical Investigation. 2015;**125**(3):926-938

[9] Dinan TG, Cryan JF. Regulation of the stress response by the gut microbiota: Implications for psychoneuroendocrinology. Psychoneuroendocrinology. 2012;**37**(9):1369-1378

[10] Foster JA, Neufeld K-AM. Gut-brain axis: How the microbiome influences anxiety and depression. Trends in Neurosciences. 2013;**36**:305-312

[11] Bercik P, Collins SM, Verdu EF. Microbes and the gut-brain axis. Neurogastroenterology and Motility. 2012;**24**(5):405-413

[12] Cebra JJ, Periwal SB, Lee G, Lee F, Shroff KE. Development and maintenance of the gut-associated lymphoid tissue (GALT): The roles of enteric bacteria and viruses. Developmental Immunology. 1998;**6**(1-2):13-18

[13] Guarner F, Malagelada JR. Gut flora in health and disease. The Lancet. 2003;**361**(9356):512-519

[14] Sudo N. Stress and gut microbiota: Does postnatal microbial colonization programs the hypothalamic-pituitary-adrenal system for stress response? International Congress Series. 2006;**1287**:350-354

[15] Harmsen HJ, Raangs GC, Degener JE, Welling GW. Extensive set of 16S rRNA-based probes for detection of bacteria in human feces. Applied and Environmental Microbiology. 2002;**68**:2982-2990

[16] O'Mahony L, McCarthy J, Kelly P, Hurley G, Luo F, Chen K, et al. *Lactobacillus* and *Bifidobacterium* in irritable bowel syndrome: Symptom responses and relationship to cytokine profiles. Gastroenterology. 2005;**128**(3):541-551

[17] Collins SM, Kassam Z, Bercik P. The adoptive transfer of behavioral phenotype via the intestinal microbiota: Experimental evidence and clinical implications. Current Opinion in Microbiology. 2013;**16**(3):240-245

[18] Maes M, Smith R, Scharpe S. The monocyte-T-lymphocyte hypothesis of major depression. Psychoneuroendocrinology. 1995;**20**:111-116

[19] Dantzer R, O'Connor J, Freund G, Johnson R, Kelley K. From inflammation to sickness and depression: When the immune system subjugates the brain. Nature Reviews Neuroscience. 2008;**9**(1):46-56

[20] Logan AC, Katzman M. Major depressive disorder: Probiotics may be an adjuvant therapy. Medical Hypotheses. 2005;**64**(3):533-538

[21] Benton D, Williams C, Brown A. Impact of consuming a milk drink containing a probiotic on mood and cognition. European Journal of Clinical Nutrition. 2007;**61**(3):355-361

[22] Mayer EA. Gut feelings: The emerging biology of gut-brain communication. Nature Reviews Neuroscience. 2011;**12**(8):453-466

[23] Mayer EA, Padua D, Tillisch K. Altered brain-gut axis in autism: Comorbidity or causative mechanisms? BioEssays. 2014;**36**(10):933-939

[24] de Vos WM, de Vos EA. Role of the intestinal microbiome in health and disease: From correlation to causation. Nutrition Reviews. 2012;**70**(1):S45-S56

[25] Park AJ, Collins J, Blennerhassett PA, Ghia JE, Verdu EF, Bercik P, et al. Altered colonic function and microbiota profile in a mouse model of chronic depression. Neurogastroenterology and Motility. 2013;**25**(9):733-e575

[26] Allen A, Dinan T, Clarke G, Cryan J. A psychology of the human brain-gut-microbiome axis. Social and Personality Psychology Compass. 2017;**11**(4):e12309

[27] Thakur A, Shakya A, Husain G, Emerald M. Gut-microbiota and mental health: Current and future perspectives. Journal of Pharmacology & Clinical Toxicology. 2014;**2**(1):1016

[28] Carabotti M, Scirocco A, Maselli MA, Severi C. The gut-brain axis: Interactions between enteric microbiota, central and enteric nervous systems. Annals of Gastroenterology. 2015;**28**(2):203-209

[29] Pistollato F, Sumalla Cano S, Elio I, Masias Vergara M, Giampieri F, Battino M. Role of gut microbiota and nutrients in amyloid formation and pathogenesis of Alzheimer disease. Nutrition Reviews. 2016;**74**(10):624-634

[30] Aziz Q, ThompsonD G. Brain-gut axis in health and disease. Gastroenterology. 1998;**114**:559-578

[31] Gillis RA, Quest JA, Pagini FD, Norman WP. Control centers in the central nervous system for regulating gastrointestinal motility. In: Schultz SG, Wood JD, Rauner BB, editors. Handbook of Physiology. Section 6. The Gastrointestinal System. New York: Oxford University; 1989. pp. 621-683

[32] Costa M, Brookes SJH. The enteric nervous system. The American Journal of Gastroenterology. 1994;**89**:S129-S137

[33] Sengupta JN, Gebhart GF. Gastrointestinal afferent fibres and sensation. In: Johnson LJ, editor. Physiology of the Gastrointestinal Tract. 3rd ed. New York: Raven; 1994. pp. 483-519

[34] Thompson DG. Extrinsic autonomic control of human gastrointestinal transit. In: Kamm MA, Lennard-Jones JE, editors. Gastrointestinal Transit: Pathophysiology and Pharmacology. Petersfield, England: Wrightson Biomedical; 1991. pp. 13-20

[35] Van Felius ID, Akkermans LMA, Bosscha K, Verheem A, Harmsen W, Visser MR, et al. Interdigestive small

bowel motility and duodenal bacterial overgrowth in experimental acute pancreatitis. Neurogastroenterology and Motility. 2003;**15**(3):267-276

[36] Rhee SH, Pothoulakis C, Mayer EA. Principles and clinical implications of the brain-gut-enteric microbiota axis. Nature Reviews Gastroenterology & Hepatology. 2009;**6**(5):306-314

[37] Chey WY, Jin HO, Lee MH, Sun SW, Lee KY. Colonic motility abnormality in patients with irritable bowel syndrome exhibiting abdominal pain and diarrhea. The American Journal of Gastroenterology. 2001;**96**(5):1499-1506

[38] Mayer EA. The neurobiology of stress and gastrointestinal disease. Gut. 2000;**47**(6):861-869

[39] Alonso C, Guilarte M, Vicario M, Ramos L, Ramadan Z, Antolín M, et al. Maladaptive intestinal epithelial responses to life stress may predispose healthy women to gut mucosal inflammation. Gastroenterology. 2008;**135**(1):163-172

[40] Jacob C, Yang PC, Darmoul D, Amadesi S, Saito T, Cottrell GS, et al. Mast cell tryptase controls paracellular permeability of the intestine. Role of protease-activated receptor 2 and beta-arrestins. The Journal of Biological Chemistry. 2005;**280**(36):31936-31948

[41] Keita AV, Soderholm JD. The intestinal barrier and its regulation by neuroimmune factors. Neurogastroenterology and Motility. 2010;**22**(7):718-733

[42] Lyte M. Microbial endocrinology in the microbiome-gut-brain axis: How bacterial production and utilization of neurochemicals influence behavior. PLoS Pathogens. 2013;**9**(11):e1003726

[43] Yang H, Stephens RL, Tache Y. TRH analogue microinjected into specific medullary nuclei stimulates gastric serotonin secretion in rats. The American Journal of Physiology. 1992;**262**(2 pt 1):G216-G222

[44] Hughes DT, Sperandio V. Inter-kingdom signalling: Communication between bacteria and their hosts. Nature Reviews Microbiology. 2008;**6**(2):111-120

[45] Lyte M, Varcoe JJ, Bailey MT. Anxiogenic effect of subclinical bacterial infection in mice in the absence of overt immune activation. Physiology & Behavior. 1998;**65**:63-68

[46] Clarke G, Stilling RM, Kennedy PJ, Stanton C, Cryan JF, Dinan TG. Minireview: Gut microbiota: The neglected endocrine organ. Molecular Endocrinology. 2014;**28**(8):221-1238

[47] Öhman L, Simren M. Pathogenesis of IBS: Role of inflammation, immunity and neuroimmune interactions. Nature Reviews Gastroenterology & Hepatology. 2010;**7**(3):163

[48] Wikoff WR et al. Metabolomics analysis reveals large effects of gut microflora on mammalian blood metabolites. Proceedings of the National Academy of Sciences of the United States of America. 2009;**106**(10):3698-3703

[49] Cho I, Blaser MJ. The human microbiome: At the interface of health and disease. Nature Reviews Genetics. 2012;**13**(4):260-270

[50] Nøhr MK, Pedersen MH, Gille A, Egerod KL, Engelstoft MS, Husted AS, et al. GPR41/FFAR3 and GPR43/FFAR2 as cosensors for short-chain fatty acids in enteroendocrine cells vs FFAR3 in enteric neurons and FFAR2 in enteric leukocytes. Endocrinology. 2013;**154**(10):3552-3564

[51] Donia MS, Fischbach MA. Small molecules from the human microbiota. Science. 2015;**349**:1254766

[52] Duszka K, Wahli W. Enteric microbiota-gut-brain axis from the perspective of nuclear receptors. International Journal of Molecular Sciences. 2018;**19**(8):2210

[53] Dumitrescu L, Popescu-Olaru I, Cozma L, Tulbă D, Hinescu ME, Ceafalan LC, et al. Oxidative stress and the microbiota-gut-brain axis. Oxidative Medicine and Cellular Longevity. 2018;**2406594**:1-11

[54] Fung TC, Olson CA, Hsiao EY. Interactions between the microbiota, immune and nervous systems in health and disease. Nature Neuroscience. 2017;**20**(2):145-155

[55] Hubbard TD, Murray IA, Bisson WH, et al. Adaptation of the human aryl hydrocarbon receptor to sense microbiota-derived indoles. Scientific Reports. 2015;**5**:12689

[56] Osadchiy V, Martin CR, Mayer EA. The gut-brain axis and the microbiome: Mechanisms and clinical implications. Clinical Gastroenterology and Hepatology. 2018;**17**:322-332. DOI: 10.1016/j.cgh.2018.10.002

[57] Hooper LV, Gordon JI. Commensal host-bacterial relationships in the gut. Science. 2001;**292**:1115-1118

[58] Clarke G, Grenham S, Scully P, Fitzgerald P, Moloney RD, Shanahan F, et al. The microbiome-gut-brain axis during early life regulates the hippocampal serotonergic system in a sex-dependent manner. Molecular Psychiatry. 2013;**18**(6):666-673

[59] Neufeld K, Kang N, Bienenstock J, Foster J. Reduced anxiety-like behavior and central neurochemical change in germ-free mice. Neurogastroenterology and Motility. 2011;**23**(3):255-264. e119

[60] Gareau M. Microbiota-gut-brain axis and cognitive function. In: Lyte M, Cryan J, editors. Microbial Endocrinology: The Microbiota-Gut-Brain Axis in Health and Disease. New York: Springer; 2014. pp. 357-371

[61] Gilbert JA, Krajmalnik-Brown R, Porazinska DL, Weiss SJ, Knight R. Toward effective probiotics for autism and other neurodevelopmental disorders. Cell. 2013;**155**(7):1446-1448

[62] Vijay-Kumar M et al. Metabolic syndrome and altered gut microbiota in mice lacking Toll-like receptor 5. Science. 2010;**328**(5975):228-231

[63] Sudo N, Chida Y, Aiba Y, et al. Postnatal microbial colonization programs the hypothalamic-pituitary-adrenal system for stress response in mice. The Journal of Physiology. 2004;**558**:263-275

[64] Cryan JF, Page ME, Lucki I. Differential behavioral effects of the antidepressants reboxetine, fluoxetine, and moclobemide in a modified forced swim test following chronic treatment. Psychopharmacology. 2005;**182**(3):335-344

[65] Schroeder FA, Lin CL, Crusio WE, et al. Antidepressant-like effects of the histone deacetylase inhibitor, sodium butyrate, in the mouse. Biological Psychiatry. 2007;**62**:55-64

[66] Rogers GB, Keating DJ, young RL, Wong M-L, Licinio J, et al. From gut dysbiosis to altered brain function and mental illness: Mechanisms and pathways. Molecular Psychiatry. 2016;**21**(6):738-748

[67] Deltheil T, Guiard BP, Cerdan J, et al. Behavioral and serotonergic consequences of decreasing or increasing hippocampus brain derived neurotrophic factor protein levels in mice. Neuropharmacology. 2008;**55**:1006-1014

[68] Rattiner LM, Davis M, Ressler KJ. Differential regulation of

brainderived neurotrophic factor transcripts during the consolidation of fear learning. Learning & Memory. 2004;**11**:727-731

[69] Drevets WC. Neuroimaging studies of mood disorders. Biological Psychiatry. 2000;**48**:813-829

[70] Bercik P, Denou E, Collins J, Jackson W, Lu J, Jury J, et al. The intestinal microbiota affect central levels of brain-derived neurotropic factor and behavior in mice. Gastroenterology. 2011;**141**: 599-609.e593

[71] Verdú EF, Bercik P, Verma-Gandhu M, et al. Specific probiotic therapy attenuates antibiotic induced visceral hypersensitivity in mice. Gut. 2006;**55**:182-190

[72] Li W, Dowd SE, Scurlock B, et al. Memory and learning behaviour in mice is temporally associated with diet-induced alterations in gut bacteria. Physiology & Behavior. 2009;**96**:557-567

[73] Bravo J, Forsythe P, Chew M, Escaravage E, Savignac H, Dinan TG, et al. Ingestion of *Lactobacillus* strain regulates emotional behavior and central GABA receptor expression in a mouse via the vagus nerve. Proceedings of the National Academy of Sciences. 2011;**108**(38):16050-16055

[74] Rodes L, Paul A, Coussa-Charley M, et al. Transit time affects the community stability of *Lactobacillus* and *Bifidobacterium* species in an in vitro model of human colonic microbiotia. Artificial Cells, Blood Substitutes, and Immobilization Biotechnology. 2011;**39**:351-356

[75] Hsiao EY, McBride SW, Hsien S, Sharon G, Hyde ER, McCue T, et al. Microbiota modulate behavioral and physiological abnormalities associated with neurodevelopmental disorders. Cell. 2013;**155**(7):1451-1463

[76] Desbonnet L, Clarke G, Shanahan F, Dinan T, Cryan J. Microbiota is essential for social development in the mouse. Molecular Psychiatry. 2014;**19**(2):146-148

[77] Clarke G, Grenham S, Scully P, Fitzgerald P, Moloney R, Shanahan F, et al. The microbiome-gut-brain axis during early life regulates the hippocampal serotonergic system in a sex-dependent manner. Molecular Psychiatry. 2013;**8**:666-673

[78] Diaz Heijtz R et al. Normal gut microbiota modulates brain development and behavior. Proceedings of the National Academy of Sciences of the United States of America. 2011;**108**(7):3047-3052

[79] Desbonnet L, Garrett L, Clarke G, Kiely B, Cryan JF, Dinan TG. Effects of the probiotic *Bifidobacterium infantis* in the maternal separation model of depression. Neuroscience. 2010;**170**(4):1179-1188

[80] de Theije CG et al. Intestinal inflammation in a murine model of autism spectrum disorders. Brain, Behavior, and Immunity. 2014;**37**:240-247

[81] Swartz TD, Duca FA, de Wouters T, Sakar Y, Covasa M. Up-regulation of intestinal type 1 taste receptor 3 sodium glucose luminal transporter-1 expression increased sucrose intake in mice lacking gut microbiota. The British Journal of Nutrition. 2012;**107**(5):621-630

[82] Duca FA, Swartz TD, Sakar Y, Covasa M. Increased oral detection, but decreased intestinal signaling for fats in mice lacking gut microbiota. PLoS One. 2012;**7**(6):e39748

[83] Crumeyrolle-Arias M et al. Absence of the gut microbiota enhances anxiety-like behavior and neuroendocrine response to acute stress

in rats. Psychoneuroendocrinology. 2014;**42**:207-217

[84] Arseneault-Breard J et al. Combination of *Lactobacillus helveticus* R0052 and *Bifidobacterium longum* R0175 reduces post-myocardial infarction depression symptoms restores intestinal permeability in a rat model. The British Journal of Nutrition. 2012;**107**(12):1793-1799

[85] Braniste V et al. The gut microbiota influences blood-brain barrier permeability in mice. Science Translational Medicine. 2014;**6**(263):263ra158

[86] Savignac HM, Kiely B, Dinan TG, Cryan JF. Bifidobacteria exert strain-specific effects on stress-related behavior and physiology in BALB/c mice. Neurogastroenterology and Motility. 2014;**26**(11):1615-1627

[87] Wong M-L, Inserra A, Lewis MD, Mastronardi CA, Kentish S, Leong L, et al. Inflammasome signaling affects anxiety- and depressive-like behavior and gut microbiome composition. Molecular Psychiatry. 2016 (in press)

[88] Bäckhed F, Ding H, Wang T, Hooper LV, Koh GY, Nagy A, et al. The gut microbiota as an environmental factor that regulates fat storage. Proceedings of the National Academy of Sciences of the United States of America. 2004;**101**(44):15718-15723

[89] Loening-Baucke V, Yamada T. Is the afferent pathway from the rectum impaired in children with chronic constipation and encopresis. Gastroenterology. 1995;**109**:397-403

[90] Ertekin C, Hansen MV, Larsson LE, Sjödahl R. Examination of the descending pathway to the external anal sphincter and pelvic floor muscles by transcranial cortical stimulation. Electroencephalography and Clinical Neurophysiology. 1990;**75**:500-510

[91] Herdmann J, Bielefeldt K, Enck P. Quantification of motor pathways to the pelvic floor in humans. The American Journal of Physiology. 1991;**260**:G720-G723

[92] Turnbull GK, Aziz Q, Hamdy S, Barlow J, Singh K, Alani S, et al. Representation of the anal sphincter on the human cerebral cortex. Gut. 1994;**35**:S30

[93] Messaoudi M, Violle N, Bisson JF, Desor D, Javelot H, Rougeot C. Beneficial psychological effects of a probiotic formulation (*Lactobacillus helveticus* R0052 and *Bifidobacterium longum* R0175) in healthy human volunteers. Gut Microbes. 2011;**2**(4):256-261

[94] Tillisch K et al. Consumption of fermented milk product with probiotic modulates brain activity. Gastroenterology. 2013;**144**(7):1394-1401

[95] Ahluwalia V et al. Enhancement of functional connectivity, working memory and inhibitory control on multi-modal brain MR imaging with Rifaximin in cirrhosis: Implications for the gut-liver-brain axis. Metabolic Brain Disease. 2014;**29**(4):1017-1025

[96] Bajaj JS et al. Modulation of the metabiome by rifaximin in patients with cirrhosis and minimal hepatic encephalopathy. PLoS One. 2013;**8**(4):e60042

[97] Tetel MJ, De Vries GJ, Melcangi RC, Panzica G, O'Mahony SM. Steroids, stress and the gut microbiome-brain axis. Journal of Neuroendocrinology. 2018;**30**(2):e12548

[98] Kassinen A, Krogius-Kurikka L, Mäkivuokko H, et al. The fecal microbiota of irritable bowel syndrome patients differs significantly from that of healthy subjects. Gastroenterology. 2007;**133**:24-33

[99] Camilleri M, Madsen K, Spiller R, Greenwood-Van Meerveld B, Verne GN. Intestinal barrier function in health and gastrointestinal disease. Neurogastroenterology and Motility. 2012;**24**(6):503-512. DOI: 10.1111/j.1365-2982.2012.01921.x

[100] Julio-Pieper M, Bravo JA, Aliaga E, Gotteland M. Review article: Intestinal barrier dysfunction and central nervous system disorders—A controversial association. Alimentary Pharmacology & Therapeutics. 2014;**40**(10):1187-1201. DOI: 10.1111/apt.12950

[101] Neufeld KA, Kang N, Bienenstock J, Foster JA. Effects of intestinal microbiota on anxiety-like behavior. Communicative & Integrative Biology. 2011;**4**(4): 492-494. DOI: 10.4161/cib.15702

[102] Gerhardt S, Mohajeri M. Changes of colonic bacterial composition in Parkinson's disease and other neurodegenerative diseases. Nutrients. 2018;**10**(6):708. DOI: 10.3390/nu10060708

[103] Liang S, Wu X, Hu X, Niu Y, Jin F. The development and tendency of depression researches: Viewed from the microbiota-gut-brain axis. Chinese Science Bulletin. 2018;**63**(20):2010-2025

[104] Booij L, van der Does AJW. Emotional processing as a predictor of symptom change: An acute tryptophan depletion study in depressed patients. European Neuropsychopharmacology. 2011;**21**:379-383

[105] Crane JD, Palanivel R, Mottillo EP, Bujak AL, Wang H, Ford RJ, et al. Inhibiting peripheral serotonin synthesis reduces obesity and metabolic dysfunction by promoting brown adipose tissue thermogenesis. Nature Medicine. 2015;**21**:166-172

[106] Jenkins T, Nguyen J, Polglaze K, Bertrand P. Influence of tryptophan and serotonin on mood and cognition with a possible role of the gut-brain axis. Nutrients. 2016;**8**(1):56. DOI: 10.3390/nu8010056

[107] Schlösser RG, Wagner G, Koch K, Dahnke R, Reichenbach JR, Sauer H. Fronto-cingulate effective connectivity in major depression: A study with fMRI and dynamic causal modeling. NeuroImage. 2008;**43**:645-655

[108] Guilloux J-P, Douillard-Guilloux G, Kota R, Wang X, Gardier A, Martinowich K, et al. Molecular evidence for BDNF- and GABA-related dysfunctions in the amygdala of female subjects with major depression. Molecular Psychiatry. 2012;**17**: 1130-1142

[109] Zheng P, ZengB ZC, Liu M, Fang Z, Xu X, Zhang X. Gut microbiome remodeling induces depressive-like behaviors through a pathway mediated by the host's metabolism. Molecular Psychiatry. 2016;**21**(6):786-796

[110] Naseribafrouei A, Hestad K, Avershina E, Sekelja M, Linløkken A, Wilson R, et al. Correlation between the human fecal microbiota and depression. Neurogastroenterology and Motility. 2014;**26**:1155-1162

[111] Theoharides TC, Doyle R. Autism, gut-blood-brain barrier, and mast cells. Journal of Clinical Psychopharmacology. 2008;**28**(5):479-483

[112] de Magistris L, Familiari V, Pascotto A, Sapone A, Frolli A, Iardino P, et al. Alterations of the intestinal barrier in patients with autism spectrum disorders and in their first-degree relatives. Journal of Pediatric Gastroenterology and Nutrition. 2010;**51**(4):418-424

[113] Crawley JN. Mouse behavioral assays relevant to the symptoms of autism. Brain Pathology. 2007;**17**:448-459

[114] Kelly J, Borre Y, O'Brien C, Patterson E, El Aidy S, Deane J, et al. Transferring the blues: Depression-associated gut microbiota induces neurobehavioural changes in the rat. Journal of Psychiatric Research. 2016;**82**:109-118

[115] Dinan TG, Cryan JF. Gut-brain axis in 2016: Brain-gut-microbiota axis—Mood, metabolism and behaviour. Nature Reviews Gastroenterology & Hepatology. 2017;**14**(2):69

[116] Kang DW, Park JG, Ilhan ZE, Wallstrom G, LaBaer J, Adams JB, et al. Reduced incidence of *Prevotella* and other fermenters in intestinal microflora of autistic children. PLoS One. 2013;**8**(7):e68322

[117] Luna RA, Oezguen N, Balderas M, Venkatachalam A, Runge JK, Versalovic J, et al. Distinct microbiome-neuroimmune signatures correlate with functional abdominal pain in children with autism spectrum disorder. Cellular and Molecular Gastroenterology and Hepatology. 2017;**3**(2):218-230

[118] Jiang C, Li G, Huang P, Liu Z, Zhao B. The gut microbiota and Alzheimer's disease. Journal of Alzheimer's Disease. 2017;**58**(1):1-15

[119] Bekkering P, Jafri I, van Overveld FJ, et al. The intricate association between gut microbiota and development of type 1, type 2 and type 3 diabetes. Expert Review of Clinical Immunology. 2013;**9**:1031-1041

[120] Zhao Y, Dua P, Lukiw WJ. Microbial sources of amyloid and relevance to amyloidogenesis and Alzheimer's disease (AD). Journal of Alzheimers Disease and Parkinsonism. 2015;**5**:177. DOI: 10.4172/2161-0460.1000177

[121] Friedland RP. Mechanisms of molecular mimicry involving the microbiota in neurodegeneration. Journal of Alzheimer's Disease. 2015;**45**:349-362

[122] Bieschke J, Russ J, Friedrich RP, et al. EGCG remodels mature a-synuclein and amyloid-beta fibrils and reduces cellular toxicity. Proceedings of the National Academy of Sciences of the United States of America. 2010;**107**:7710-7715

[123] Lozupone C, Knight R. UniFrac: A new phylogenetic method for comparing microbial communities. Applied and Environmental Microbiology. 2005;**71**(12):8228-8235

[124] Agarwala S, Naik B, Ramachandra NB. Diversity of gut microbiota in autism reveals differential abundance of *Prevotella* and *Akkermansia* species. 2018

[125] Chapelet G, Leclair-Visonneau L, Clairembault T, Neunlist M, Derkinderen P. Can the gut be the missing piece in uncovering PD pathogenesis? Parkinsonism & Related Disorders. 2018 (accepted manuscript)

[126] Rodríguez-Violante M, Zerón-Martínez R, Cervantes-Arriaga A, Corona T. Who can diagnose Parkinson's disease first? Role of pre-motor symptoms. Archives of Medical Research. 2017;**48**(3):221-227

[127] Cova I, Priori A. Diagnostic biomarkers for Parkinson's disease at a glance: Where are we? Journal of Neural Transmission. 2018;**125**(10):1417-1432

[128] Heintz-Buschart A, Pandey U, Wicke T, Sixel-Döring F, Janzen A, Sittig-Wiegand E, et al. The nasal and gut microbiome in Parkinson's disease and idiopathic rapid eye movement sleep behavior disorder. Movement Disorders. 2018;**33**(1):88-98

[129] Leclair-Visonneau L, Clairembault T, Coron E, Le Dily S, Vavasseur F, Dalichampt M, et al. REM sleep

behavior disorder is related to enteric neuropathology in Parkinson disease. Neurology. 2017;**89**:1612-1618

[130] Knudsen K, Fedorova TD, Hansen AK, Sommerauer M, Otto M, Svendsen KB, et al. In-vivo staging of pathology in REM sleep behaviour disorder: A multimodality imaging case-control study. Lancet Neurology. 2018;**17**:618-628

[131] Peisl L, Schymanski EL, Wilmes P. Dark matter in host-microbiome metabolomics: Tackling the unknowns—A review. Analytica Chimica Acta. 2018;**1037**:13-27

[132] Scheperjans F, Derkinderen P, Borghammer P. The gut and Parkinson's disease: Hype or Hope? Journal of Parkinson's Disease. 2018;**8**(s1):S31-S39

[133] Knudsen K, Fedorova TD, Hansen AK, Sommerauer M, Haase AM, Svendsen KB, et al. Objective intestinal function in patients with idiopathic REM sleep behavior disorder. Parkinsonism & Related Disorders. 2018 (accepted manuscript)

[134] Sudo N. Microbiome, HPA Axis and production of endocrine hormones in the gut. Advances in Experimental Medicine and Biology. 2014;**817**:177-194

[135] Kim S, Kim H, Yim YS, Ha S, Atarashi K, Tan TGL, et al. Maternal gut bacteria promote neurodevelopmental abnormalities in mouse offspring. Nature. 2017;**549**:528-532

[136] Bien-Ly N, Watts RJ. The blood-brain Barrier's gut check. Science Translational Medicine. 2014;**6**:263fs46

[137] Hoban AE, Stilling R, Ryan F, Shanahan F, Dinan T, Claesson M, et al. Regulation of prefrontal cortex myelination by the microbiota. Translational Psychiatry. 2016;**6**(4):e774. DOI: 10.1038/tp.2016.42

[138] Neufeld KA, Foster JA. Effects of gut microbiota on the brain: Implications for psychiatry. Journal of Psychiatry & Neuroscience. 2009;**34**:230-231

[139] Jiang H, Ling Z, Zhang Y, Mao H, Ma Z, Yin Y, et al. Altered fecal microbiota composition in patients with major depressive disorder. Brain, Behavior, and Immunity. 2015;**48**:186-194

[140] Sherman PM. Probiotics and lactose maldigestion. Canadian Journal of Gastroenterology. 2004;**18**(2):81-82

[141] Ledochowski M, Widner B, Bair H, Probst T, Fuchs D. Fructose- and sorbitol-reduced diet improves mood and gastrointestinal disturbances in fructose malabsorbers. Scandinavian Journal of Gastroenterology. 2000;**35**(10):1048-1052

[142] Savignac H, Tramullas M, Kiely B, Dinan T, Cryan J. Bifidobacteria modulate cognitive processes in an anxious mouse strain. Behavioural Brain Research. 2015;**287**:59-72

[143] Allen A, Hutch W, Borre Y, Kennedy P, Temko A, Boylan G, et al. *Bifidobacterium longum* 1714 as a translational psychobiotic: Modulation of stress, electrophysiology and neurocognition in healthy volunteers. Translational Psychiatry. 2016;**6**(11):e939

[144] Claesson M, Jeffery I, Conde S, Power S, O'Connor E, Cusack S, et al. Gut microbiota composition correlates with diet and health in the elderly. Nature. 2012;**488**(7410):178-184

[145] Reuter J, Spacek D, Snyder M. High-throughput sequencing technologies. Molecular Cell. 2015;**58**(4):586-597

[146] Levy RL, Olden KW, Naliboff BD, Bradley LA, Francisconi C, Drossman DA,

et al. Psychosocial aspects of the functional gastrointestinal disorders. Gastroenterology. 2006;**130**(5):1447-1458

[147] Whitehead WE. Twin studies used to prove that the comorbidity of major depressive disorder with IBS is NOT influenced by heredity. The American Journal of Gastroenterology. 2007;**102**(10):2230-2231

[148] Canavan C, West J, Card T. The epidemiology of irritable bowel syndrome. Clinical Epidemiology. 2014;**6**:71-80

[149] Mayer EA, Tillisch K, Bradesi S. Review article: Modulation of the braingut axis as a therapeutic approach in gastrointestinal disease. Alimentary Pharmacology and Therapeutics. 2006;**24**(6):919-933

[150] Sampson TR, Debelius JW, Thron T, Janssen S, Shastri GG, Ilhan ZE, et al. Gut microbiota regulate motor deficits and neuroinflammation in a model of Parkinson's disease. Cell. 2016;**167**: 1469-1480.e12

[151] Papakostas GI, Thase ME, Fava M, Nelson JC, Shelton RC. Are antidepressant drugs that combine serotonergic and noradrenergic mechanisms of action more effective than the selective serotonin reuptake inhibitors in treating major depressive disorder? A metaanalysis of studies of newer agents. Biological Psychiatry. 2007;**62**(11):1217-1227

[152] Kennedy SH. A review of antidepressant treatments today. European Neuropsychopharmacology. 2006;**16**(5):S619-S623

Section 2

Behavioral Neuroscience in Developmental Age - Examples of Hot Topics in Healthy and Pathological Subjects

Chapter 4

Challenges for Behavioral Neuroscience: Prenatal, Postnatal, and Social Factors

Götz Egloff and Dragana Djordjevic

Abstract

Behavioral medicine has neglected social aspects for a long time. In the pre- and postnatal context, these are especially important, as parental competencies in the relational objects of the infant may be compromised by both inner and outer factors, thus potentially compromising the infant's psychic development. The findings on pre- and postnatal stages of human development have shown that early psychosocial interventions can help out to some extent. Approaches for parents, which have mainly evolved from the findings of psychoanalysis and mother-infant research, must be augmented by a social perspective, just like postnatal concepts have been augmented by prenatal intervention approaches. The latter reach from how parents-to-be can be prepared for parenthood to how to support attachment and relation in infants, toddlers, and older children. Scientific behavioral reasoning, augmented by subjectivity- and objectivity-related concepts, provides a framework to work with, so that potential deprivation can be faced seriously. Intervention approaches focusing on bonding, on relational issues, and on educational practices are introduced, covering the most important time spans of psychic development from the mother-unborn period to the mother-infant period.

Keywords: prenatal, postnatal, social, interventions, mother-infant research

1. Introduction

Psychoanalysis and infant mental health research offer a large amount of knowledge about human development, pathology and interventions, which can partially be grounded in the findings of neuropsychoanalysis [1, 2]. These findings connect to what Stierlin conceptualized as relational individuation, or co-individuation ("Bezogene Individuation": the principle that "the higher the level of individuation in a family member, family, or group is, the higher the level of personal relating becomes, and is required at the same time") [3], a concept that originally aimed at psychic identities of members of a family system. The concept generally supports the principle of socialization modes in Lloyd deMause's approach of psychogenetic personality development [4, 5]. The psychogenetic personality concept illuminates the modes of manifestation of transgenerational psychodynamics, and even takes into account physiologically based premature birth in humans [6]. The concept hints at the mutual interaction of individual and societal development [7], which will show in subjectivity and in socialization modes. Illuminating their organic

IntechOpen

substructure might be one of the challenges of future behavioral neuroscience, an interdisciplinary exchange of concepts and of mutual impregnation, the aim of future scientific cooperation. The question of how to bring together brain, mind, and the social will be one of the difficult tasks.

In the course of recent infant mental health research, fetal brain development has been examined from a bio-psychosocial perspective [8], as has been by Roth [9] from a neuroscientific one. In his depiction of prenatal and early postnatal processes of brain development, three levels in the limbic system correlate with temperament, with early experiencing, and with subsequent socialization, of which the latter may be responsive to compensatory intervention. Psyche, in the neuroscientific perspective, is strictly related to brain physiology, a controversial [10] still worthwhile approach, since it has been shown that early experiences will influence brain function and structure in humans [11]. What has mostly been accepted is that the concept of subject autonomy is generally challenged owing to Freud's observation, "Der Reflexvorgang bleibt das Vorbild auch aller psychischen Leistung (The reflex act remains the type of every psychic activity as well)," [12] which he stated to put psychic mechanisms in connection with automatic reflex processing, in order to emphasize the predominance of unconscious psychic processes. Around 100 years later, the findings about intuitive responses being in middle position between innate reflex behavior and seemingly more "rational" behavior, have been brought up thanks to video microanalyses of dyadic interaction of infants and parents. In this context, parental competencies are referred to as intuitive competencies [13–15]. These are elicited within a time frame of 200–600 ms. Not only mothers, also fathers, children, and other relational objects have been observed to show these; they are universal and to be found in persons of any age, any gender, and in any culture [16].

Spitz, in the 1960s, observed that the physical presence of the mother, i.e., of one relational object, is the basic precondition for successful infant mental development [17, 18]. Severe social deprivation in hospitalized and institutionalized children, which grew up without responses to their needs showed compromised development in many aspects [19, 20]. By now, diagnostic approaches and options of treatment of infants and toddlers even encompass a psychodynamic concept [21], focusing on conflict, structure, and relation perspectives, thus paving the way for developing a rather focus-oriented treatment approach. This will probably be used more frequently in the future, just like operationalized psychodynamics in OPD-2 has increasingly been used in studies of the last years [22]. Although operationalized psychodynamics has widely improved the clinical view of human development, misconceptions have not been avoided: specific cultural and social influences on the infant's development are still grossly neglected. Socioeconomic factors on mothers' sensitivity and on family functioning have only begun to be examined [23].

2. Anthropological basics and mother-infant research

The intrauterine development of the cerebral cortex occurs in exact stages. Each developmental step is a vulnerable period, which is sensitive to insults rendering the brain susceptible to structural malformations and functional impairments [8]. Neurogenesis shows billions of neurons being produced during the development of the central nervous system. It mainly occurs at the inner edge of the neural tube wall, later ventricles, and spinal cord. Cell division begins once the neural tube has closed at 4–5 weeks after conception, which is 6–7 weeks of gestation. Most neurons are formed at 12–18 weeks of gestation. Around 200 billion neurons are produced in the human brain, and 40 billion in the neocortex alone, of which the half are eliminated during the maturation process, resulting in a final number of 100 billion neurons at 40 weeks in full-term infants. Maternal stress during the first trimester has been linked to an increased risk of pathology, suggesting that the expression of genes in early fetal life is influenced by external factors, leading to behavioral and cognitive malfunction or to psychiatric disorder like schizophrenia [8]. Stress-induced reduction of neurons in late fetal life is probably associated with increased damage of neurons. Adding to it, the conspicuous findings on correlations of maternal mental disorders in pregnancy to the child's subsequent psychic development can be examined from different perspectives, as can psychic development within a broader context.

From a psychoanalytic perspective, there is a perinatal constant of originary separation as inscription of lack within the ego. It is a separation of the ego from the developing subject through "objet petit a." The object, the so-called other, is the object-cause of desire. It is the driving force, which makes the subject seek something, organically mirrored in the mesocortical and mesolimbic seeking systems of the frontal lobe. The subject in encountering the object experiences entering the Real beyond symbolization. If it was not for physiological prematurity in humans [24], one might argue that human seeking is merely for reasons of expansion, or exploring. Still, it is originary separation adding to physiological prematurity, which seems to induce primary "homelessness" in *Homo sapiens* [7]. The subject comes to exist through seeking only.

While German pioneer of psychosomatics Thure von Uexkuell gave point to the tuning of inner and outer world in animal life, in humans the relation to nature is flawed, or altered ("altéré"). It is altered, Lacan noticed, "through a certain dehiscence (*déhiscence*) of the organism internal, through a primordial discord (*discorde*) (...), as is shown in the signs of discontent and in physical incoordination in the first months of the newborn. The objective rationale of the anatomical imperfectness of the pyramidal system (...) confirms this view, which we formulate as obtaining true specific prematurity of birth in humans" [25]. Along such an anthropological constant, it should be common ground to assume biological and sociopsychological aspects to be relevant to human development.

The biological aspect refers to instinctual life in connection with separation. Anxiety is the most basic of experiences and can be reactualized at any time. Such a reactualization of anxiety figures in anxiety of the cut ("coupure"): it is in cutting, dissociation, separation [26], which is first and foremost in birth, and then in castration and punishment. Although the latter belong to the sphere of the Symbolic, they actualize the first, stemming from the sphere of the Imaginary. As Catherine Malabou points out, death is prefigured in castration. Castration anxiety does not primarily represent the loss of a specific object but rather the indeterminate threat of separation, of a cut. In connection with repetition compulsion and the "fort/da" game in "Beyond the Pleasure Principle" [27], the anticipation of separation from the self is a primal motive ("Ur-Angst"). Any trauma experienced is in terms of such psychodynamics, namely since probably "all events—even 'real' or traumatic events—ultimately occur at the heart of the psyche's separation from itself (...)" [26].

The psychological aspect refers to symbolization through employing language in human development after the early mother-infant stage has been passed through. At that stage, it is about basic trust ("Ur-vertauen") in order to overcome mechanistic thinking ("pensée opératoire") and alexithymia in the infant, and the Imaginary has provided space for the infant to develop. Symbolization will increasingly enter the Imaginary, over-writing the experiences the infant has made before. Ludwig Janus has called attention to the concept of transcription, or transliteration ("Umschrift"), occurring from one developmental stage to another, which Freud in a letter to Wilhelm

Fliess remarked in 1896 [28]. However, the failing of symbolical over-writing, i.e., of restructuring or rewriting of the Imaginary, can be a marker of psychosis [29], or, on the imaginary-organic pole, of asymbolic conversion [30]. In "The Interpretation of Dreams" [12], Freud describes dreams as expressions of the primary process, in which wish fulfillment is executed. Thus, dream, delusion, and confabulation and other psychosis-like disorders of thinking can be viewed as working temporarily in lieu of the demands of the frontal lobe control system.

From a psychoanalytic perspective, disorder is the result of quite a normal struggle for conflict solution in differing gradations of primary, i.e., preverbal, processual thinking, and secondary, i.e., verbal, processual thinking. Pathology can be read in gradations of normalcy. Outer stressors can trigger a reactualization of preverbal, i.e., pre- and postnatal affects [7]. These encompass many factors contributing to compromise conflict solution. Disorder will take the very gradation the subject is susceptible of, only to produce as little conflictive tension as possible. At preverbal stages, disorder will mainly be body disorder; conversion can take place before any symbolization is possible.

At this point, the seemingly societal decline of symbolic references might make any structural framing, i.e., inner positioning, difficult to achieve [29]. Inner positioning can be taken as being connected to outer positioning in its literal sense: as an example, the ancient Greek "polis" would be a place of enabling positioning for—a few select—people to grow into thinkers like Plato and Aristotle. For them, it would provide space and structure to developing thought and concept. A frame would be provided in which personal relations could grow into becoming the background of successful development [31]. In contrast, today's ever-existing interpellation to people manifests in a very concrete societal trend of commodified relations calling for even less inner positioning, adding to, and retroacting on, the withdrawal of sustained societal structure, rendering more and more impossible people participate in major social achievements [32]. Such a trend might also compromise psychic functioning of parents and mothers-to-be severely. It might be the phenomenon of "new morbidity" in infants, which is in the trend toward early functional and psychic disorders and toward chronification, closely associated with this societal trend. At any rate, biological and psychological aspects of human development show humans to be prone to dysfunctional internal conflict processing, probably even more when obscure personality traits seem to be promoted [33], while virtual media foster the loss of sense of reality [34]. Such depravation can be seen in the phenomenon that present-day western societies increasingly call for behavioral and experiential conformity: changing its character, the issue of diversity becomes an interpellation of conformity. Depravation retroacting on poor psychic structure in people might prevail for generations [15], especially when a societal mode of too much freedom in some areas and too much restraint in others takes effect. It would be worthwhile examining how the organic substructure of psychological functioning and societal superstructure are intertwined, and how the "culture of commodification" [35] affects the mother-infant relationship. Looking at research on mirror neurons [36], it is not out of the question that such processes affect subject development on a macrolevel beyond the microlevel of mother-infant relations.

On a microlevel, especially postpartum depression in mothers has been the subject of extensive research investigations [37–39]. As is widely known by now, in interaction with their newborns depressive mothers show decreased responsiveness, increased passivity and/or intrusiveness, increased withdrawal, and decreased expression of positive emotions, and they tend to regulate the effects of their newborns in an insufficient way. Moreover, Papoušek and von Hofacker [40] have generally pointed out the connection between psychopathology traits in mothers and maladaptive patterns in their intuitive competencies. Correlations of maternal mental disorders in pregnancy to the child's subsequent psychic development [41] are often conspicuous, yet by no means automatic. Such are certainly individual and can be influenced. Infant mental health observations could show slight but distinct negative influences of infant crying and sleeping problems on the child's subsequent social development [42]; infants' regulatory problems contribute substantially to external and internal psychic problems in early childhood [43]. Adding to recent behavioral oxytocin research [44] of the human "attachment system," from a perspective of behavioral neuroscience, it would be worthwhile exploring the testosterone-perspective of the human "lust system," and the dopamine-perspective of the human "love system."

Familial strain of different kinds can lead to dysfunctional relational patterns; missing or inadequate internal educational models in parents can have a similar effect [45]. As to intuitive parental competencies, it has become evident that intrapsychic and interpersonal factors can compromise the expression of these. Likewise, it must be assumed that the level of expressing such competencies might be dependent on social factors. This issue has widely gone unrecognized [15, 46]. Social factors viewed from a microperspective give way to a questionably individualized concept in which societal motion, e.g., toward fragmentation and irrationality through anomic tendencies, is neglected. It should not be surprising to see irrationality increase with too many choices [47]; any compromising of the formation of psychic function will lead to people's attempts at escaping mentalization. Given mentalization is the key to at least some of parental competencies issues, more are still pending; e.g., the capability of executing ego-functions may have developed in an individual, but may not be expressed. At any rate, in a perspective of mentalization as basic ego-function, such capability is a precondition of role-taking and changing of perspectives. Empathy corresponds with this function and is often missing, especially in somatoform disorder. In practice, somatoform disorders are often diagnosed as functional syndromes [48]; i.e., somatization shows in body disorder. Alexithymia often accompanies somatoform disorder; it should also be viewed in a context of societally induced personality issues.

Parents´ cause attributions often reveal such a connection. In general somatoform disorder, from both older children and parents, psychosocial cause attributions are more often the case than in, e.g., asthma bronchiale to which rather genetic, external, and somatic causes are attributed [49]. Also, there are only moderate matches of subjective disorder beliefs in older children and parents: preframed attribution questionnaires generate higher scores of matches than half-open qualitative explorations do [50]. Generally, high diversity in parents' knowledge and cause attributions of their children's symptoms [51] invites to improve communication on many levels. Although it is obvious that pathological personality traits are associated with the ability to understand emotional states of others [52], social cognition aspects, of which mentalization is mostly in focus, are on a microperspective of family interaction. Although subscribing to a psychodynamic perspective, hereby only a small aspect is examined. In case that identity issues play a significant role, identification presupposes an original to identify by [53]. It has to meet requirements of highly differing concern [54] and has to do with subjective experience of identity [55].

3. Development and pathology

Fedor-Freybergh, from a prenatal viewpoint, has rehearsed the problematic nature of increasing discontinuities emerging from social destabilization reaching back to early prenatal traces of memory [56, 57]. The general message seems to be that through the processes of neuronal migration, organization, wiring, myelination, shaping and eliminating of excess neurons [8], even earliest information is sustained. Still, an approach of the earlier the better in several aspects of intervention has not yet been fully realized, as can be derived from the findings of epigenetics and fetal programming [58].

In prenatal stages of increased neuronal plasticity, milieu factors influence protein synthesis and program reference input in biological systems such as the HPA axis. While early postnatal epigenetic alterations are still partially influenceable [59, 60], the Barker hypothesis [61–63] postulates highly probable influences from the fetal period on cardio-metabolic functioning [62] and on brain functioning [64]. Some pregnancy-associated disorders have shown to connect to fetal experiencing [65], which also hints at the fetal period to be highly important for psychosomatic development. At least, it can be said to be responsible for the development of an archetypal mode of bonding and ambivalence ("Urbindung" and "Urambivalenz"). Taking into account that regulatory disorders in infants are obviously correlated with insufficient dealings within the family system, especially the family but also institutional surroundings of early childhood like kindergarten and preschool play a significant role in influencing personality. Research findings on regulatory disorders [66–70] provide dyadic insights but do not tend to regard triads [71], let alone setting, context, or background [72]; that is why many findings of attachment research [73] need to be augmented by a more panoramic view of relations. Also, an intergenerational perspective of trauma impact [74] carries weight since it provides vertical insight into modes of re-traumatization.

The pivotal role in human developmental pathology is certainly played by violence, as it shows in externalized action with huge destructive potential. Individuals with violent behavior inflict injuries on others, either physically, psychically, or both. Individual; i.e., subjective violent behavior, as social scientists like Hurrelmann [75] have shown, is mainly to be understood as generated by intrapsychic, interpersonal and social conflicts. Still, even an obvious inclination to aggression must not be assessed pathological in general; aggression encompasses a zestful constituent part [76]; it goes heavy on libido, i.e., on the dopaminergic system. That is why violence must not be confused with aggression in the shape of expansion and initiative, which belong to the individual developmental process. In contrast, violence as a mode of destructive aggression will have to undergo a transformation into pro-social modes before it is realized. As is often the case, etiopathology of psychic disorders can only partially be traced back [77]. Yet, concepts of phenomenology like, e.g., pathogenetic situation [10, 39], can reasonably be applied, and diverse traits of complex trauma can be observed out of which violence emerges [78]. Traumatized children have problems with changing perspectives since persisting stress from complex trauma has severely compromised their modes of experiencing, adding to lifelong trauma-associated conditions like dissociation [78, 79]. Presently, a phenotypical similarity in dissociation and severe psychopathology like schizophrenia is being discussed [80].

It is obvious that high levels of interdisciplinary exchange will be necessary to meet the challenges of brain, mind, and social factors (cp. **Figure 1**). In order to conceptualize further research on their intertwining, subjectivity formation and social objectivity have to be differentiated. The following concepts are thus not along the differentiation of subjective and objective aspect in dual-aspect monism as in the conception of Kessler et al. [81] but describe the subject in a grid of collective predisposition into which it has to develop.

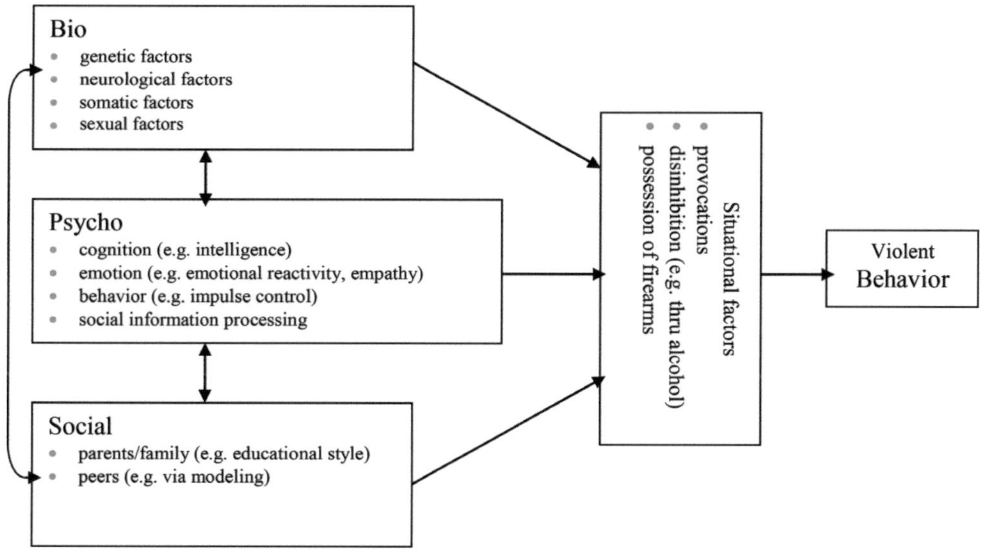

Figure 1.
Bio-psychosocial factors model of violent behavior, modified after Schick 2017.

3.1 Subjectivity formation

The mirror stage in the infant's development provides no coherent experience of the image in the mirror. Anamorphic as it is, it tends to convey rather fragmented than coherent aspects of the personality-to-be. That is why Lacan considered coherence an illusion, also owing to the fact that infantile dependence and helplessness are not conveyed in the mirror image. In referring to physiological prematurity, Lacan is close to Otto Rank's concept in which the whole self ("Total-Ich") precedes the partial self ("Partial-Ich"). Anything which is postnatal will only remain partial. Along birth, any wholeness will inevitably be lost: this is what humans will have to accept in life [82]. Here, we have a deceptive case of anthropology: there is not any totality possible. Infantile identification with the mirror image brings about alienation, or alteration, to the emerging subject, as well as dehiscence and discord—seemingly biological yet a specifically human feature [83]. Basic vulnerability stems from this stage; it can shake the infant when it is confronted with outer objects. Any identification, e.g., with parents, siblings, or teachers bears refractions.

When the mother reflects the infant, the infant creates an imaginary space through projecting his own reflected bodily self [30]. It is eventually connected to the fantasy, or anticipation, of separation by a cut. This phenomenon is linked to the illusion of coherence, which provides stability; on the other hand, there is a subversion-proneness due to an inherent amount of fictitiousness and externality within the developmental process. In the course, the outer world is perceived more coherent, more indisputable than it really is. More often than not, those objects out there are experienced as identifiable egos having unity, permanence and, first of all, substance. But those objects generally comprise a fair share of ourselves, which we tend to have abdicated ambivalence and fragmentation: after all, we wonder why those objects are that fragile. So, imaginary coherence provides people with anxiety too. The earliest developmental stages, pre- and postnatal, are gateways to imaginary formations of ideals via identification and reproduction of social roles. Taking on societal relations that begin at this point, the subject remaining is prone to ideological indoctrination. Social environment might fill the subjects' fantasies at worst distracting the subject from recognizing reality, eventually leading to escapism [84]. The infant's bewilderedness at that stage makes for irritation, and for readiness to fetch interpellation.

Violent behavior is to be called subjective violence, as it is clearly visible and shows in acts of crime. Yet, the location of subjective conflict is not necessarily identical with the location of expressed violence. Children often enact at-home-conflicts in school or kindergarten. Experiences of victimization and conflict may be brought back home, leading to aggressive behavior, e.g., in sibling or in parent interaction. At any rate, violent behavior may be used as a personal solution within a given structure, thus subjective acting manifests as violent acting. What is known is that in families with high psychic dysfunctionality parents are not capable of taking enough care of their children, either physically or psychically. Subjectively violent individuals often seem to have such a background [85], and they have often been victims of violence themselves [86]. Sometimes there has been a lack of attachment in mother-infant-relations existing from birth onwards, or there are disorders of early attachment that have developed in the infant's first year of age, or different sorts of subjective psychopathology in parents affect the infant's emotional development. Still, social status and the status of societal development may compromise psychic competencies, as can be concluded from very different research perspectives [33, 34, 87]. Dysfunctional and noncoherent educational practices in some families, which can puzzle and disturb children and direct their development toward dysfunctional modes of behavior may even be amplified by the loss of societal structure; at least it may disturb families in developing consistent educational modes [15, 46]. Some findings on subjective violence indicate an early lack of empathy in children, a lack of impulse control, and a lack of anger management in connection with early deprivation phenomena. Deviant behavior in the shape of criminal behavior can be viewed as developmental pathology, especially if lack of empathy or lack of emotional reactivity [88] can be diagnosed. Even when in offenders lack of reflective functioning [89] seems to be the key to their violent acting, and their experiences of abuse and violent behavior can be linked to their lack of individual mentalization [90], an important role in socialization must also be seen in educational institutions' repressive force, which mostly will not support empathy but competition. Competition may not be bad, still empathy needs to be supported as levels of empathy indicate the levels of pro-social behavior [91]. Moreover, any subjective behavior can be viewed as a solution-type compromise that is workable on a personal level and is due to the dialectics of acquiescence and resistance in the process of subjectivity formation. Even when such behavior may only be one among several psychic solutions of the individual, it cannot be surprising when some children react violently according to their personal biographic experiencing (cp. **Figure 2**)—which would be a long-term and somewhat functional mode of behavior [92].

3.2 Social objectivity

While zestful aggression makes for what can be called anthropology of the political [76] that does not deny subjective libidinous aggression aspects, violence must be viewed from a perspective of multifactorial subjective and objective connection. Objective violence is to be differentiated from the subjective kind [93, 94]. Contrary to subjective violence, which is committed by individuals and groups, objective violence emerges through objective reality itself; it is systemic, anonymous violence that is seemingly without reason but conceptual, more uncanny than direct precapitalist socioideological violence, which could be imputed to individuals' intentions [94]. Objective violence stems from the generated frame in which people exist and act. It is the societal background in which ideology evolves in the

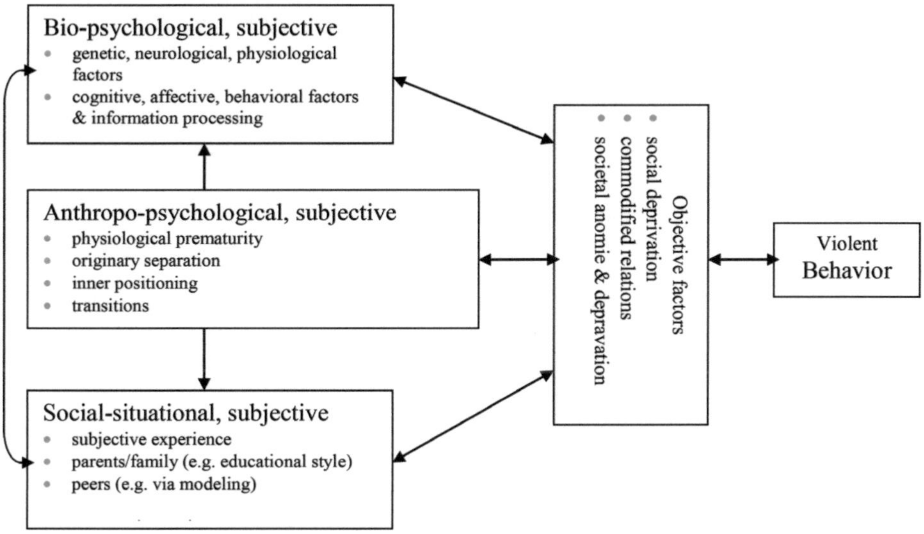

Figure 2.
Subjective and objective factors model of violent behavior.

subject. Individuals expressing subjective violence in this context have to be viewed being subjectively and objectively motivated. From an objective perspective, violent acts might be an attempt at realizing representation [32]; from a subjective perspective that would be the wish for reality—which turns out to be second reality [32, 34]. That would be a matter of substructure hitting upon superstructure, as is brain hitting upon societal commodification demands [95]. The contours of society are not only shaped by continuous interpellation through societal systems of economics and politics but the seemingly smooth functioning of society is at the same time obliged to generate outbursts of individual, i.e., subjective violence. What may be conceptualized as personal shortcomings in individuals can also be traced back to objective violent structure characterized by societal depravation. What may appear as solely internal conflicts the subject has to solve seems co-determined by the ideological structure that dominates their surroundings.

As there are cultural differences in societies, which are said to be quite similar—a mundane example is that Americans show higher scores of body image dissatisfaction than Italians [96]—it is that what may look like internal processes only should also be viewed as the result of internalized societal relations of which an individually processed relation of the subject to their surroundings is formed. This relation may either, more or less, remain on a fantasmatic level tending to repress reality, or develop toward a rather realistic level. More than enough, human readiness for projective processing [97], i.e., for fantasmatic modes of creating personal reality, is an anthropological constant, which seems due to physiological prematurity in humans in the course of evolution. Adding severely to it, subject-object-differentiation nowadays is increasingly blurred owing to the loss of representation in virtualized surroundings [32]. That is why it is not possible to retrieve authenticity, if ever there was one. As people tend to hang on to the concept of authenticity especially in highly virtualized surroundings, "one always wishes to see the other act naturally, but this eludes him and thus becomes an object of fetish and intrigue" [55].

Fundamental issues of identification and representation still go unresolved [53]. Societal motion may seem detached from individual action at first but is not. It has strong effects on everyday dealings. Objective societal structure, at least in Western Europe, is currently dominated by high degrees of personal freedom and its concurrent, restraint, at the same time. The shibboleth of absolving societal structure from its responsibility of taking effect on human living conditions [98] promotes such a motion. Instead, bio-psychosocial environment viewed as a result of early interaction combined with societal interpellation hitting upon organic substructure provides a reasonable framework to work with. Certainly, given the inevitable entanglement of the individual in the socialization process confronting multiple determinants [71], the question must be raised whether or not a subject can be a subject undamaged at all [99]. Peter Zima ascertains the subject to be inherently pending between rejection and indispensability, between subjugation and freedom [100]. Still, as Resch and Parzer point out, it is not subjective realities and interpretations that will prevail but phenomena like death, pain, and poverty. Such phenomena cannot be misread, cannot be reframed [101]. They belong to objectivity. Only some of deprivation phenomena are man-made, while others are not.

4. Pre- and postnatal interventions

4.1 Findings

Early communication influences development and learning processes in children [102]. On a microlevel, Papoušek has described the significance of communicative acting for early emotional relatedness [103]. On a macrolevel, phylogenetic human development concepts have augmented ontogenetic aspects of prenatal and perinatal development, broadening the concepts of postnatal development. Examples of how prenatal psychosomatic factors in mothers-to-be can affect their experiencing and retroact on gestation and delivery have often shown congruence, all the more those with a focus on imminent preterm delivery [65, 104], which is an issue with high significance as preterm infants require special treatment [105]. The issue of neurodevelopmental outcome in preterm newborns is still highly problematic and connected with new morbidity [70, 106–110], therefore new approaches in neonatal intensive care units have been developed and implemented [14, 70]. Recent findings that, e.g., preterm delivery correlates with infant eating disorders [111] should not be surprising; other findings indicate prenatal and perinatal factors in new morbidity [58]. Today's zeitgeist has only begun to be examined: it tends to favor noncommitment [101], pointing to the connection of new morbidity with societal motion [97].

Although for decades there have been efforts reaching out to prenatal aspects of mental health [112], structured programs are relatively new in Germany. As to overall parenting and early childhood, interdisciplinary and cross-cultural collaborations have emerged [113, 114]. On an individual level of childhood education, structured programs have successfully been established in many educational institutions in Germany [115–117]. The majority of early intervention programs available have mostly been adaptations from the USA. Comparisons between USA and German programs have proven to be difficult due to structural differences in health services [118]. In Germany, they mostly focus on the mother-infant-relationship [114], when a focus on postpartum depression and anxiety would be equally relevant. Recent meta-analyses show that programs starting during pregnancy were evaluated as the best when they had a high frequency of home visits [119, 120]. There are findings of advantages of close and personal relating to one another, which comes close to a therapeutic setting. Moreover, maternal symptom burden was relieved the most in a setting with psychotherapeutic elements established for mothers having to cope with preterm delivery [121]. Generally, maternal symptom burden relief has been the most observed effect in programs while there were only small effects in interventions on maternal competencies re-enforcement.

Also, only small effects on child development have been observed, and these have been lower and more heterogeneous than the effects on mothers-to-be. Then again, having more than 20 sessions has proven helpful for the infant's physical development [119].

The early intervention approaches depicted below are not supposed to be therapy for mothers and infants. Instead, these are psychodynamics-oriented programs and concepts, which focus on potentially significant topics in pre- and postnatal stages of development. They have evolved from many of the findings above and have purposely been designed to support mental health of parents and children: from the unborn during pregnancy to the newborn and after, and to parenting in general. The programs take care of the microlevel of inner family issues. Here they are presented in order of diachronic developmental aspects reaching from prenatal to postnatal development.

4.1.1 "Mutter-Kind-Bindungsanalyse" (mother-infant bonding analysis)

Mother-infant bonding analysis [122, 123] is a procedure of accompanying women in pregnancy enabling them to get in contact with their unborn; an approach for which Phyllis Klaus's work paved the way [124]. It is not a structured program in the narrow sense of the word but a fairly structured interventional sequence of individual sessions. By these, early before delivery first steps of building a relation between mothers-to-be and their unborn are encouraged. Through relaxation on a couch, women focus on their perception of signals from the unborn. These will show in the shape of emotions, images, thoughts, and fantasies on a so-called "inner screen," which both unborn and mother are related to. This communicative channel can be seen as "umbilical cord" of psyche, enabling a dialog, which is supposed to promote the intrauterine development of the unborn. The bonding analyst will support mothers-to-be get in contact with the unborn by encouraging them, by interpreting, and by helping to overcome blockades if necessary. Twenty to thirty sessions during the second half of pregnancy are usually taken, that is from twentieth to fortieth gestational week. Exactly this time frame is known as the unborn's highest brain sensitivity and vulnerability period [125]. The history of mother-infant bonding analysis goes back to the early 1990s when Budapest-based Jenö Raffai recognized in his work with patients the importance of the prenatal mother-unborn-relationship for the infant's and the adult's further development. Together with the Hungarian psychoanalyst György Hidas, he conceptualized a research and treatment method that developed into bonding analysis. Especially the focus on children's personality development through the well-being of mothers in pregnancy and birth-giving might serve as the prenatal reference to autobiographical memory [126].

4.1.2 "SAFE"

The structured program "SAFE"—Sichere Ausbildung für Eltern (Secure Education for Parents) [127] aims at what is best for mothers in pregnancy, during delivery, and in parenting issues. The main issue of the program is to avoid transferring of traumatic childhood experiences toward the infant. "SAFE" helps parents-to-be develop confidence in dealings with the infant. As early as in pregnancy they learn to recognize and react appropriately to the signals the infant shows. This is helpful in developing a secure mode of attachment in infants since securely attached infants show more capability of empathy, are more creative, and are more capable of cognitive processing, as well as they search easier for help when needed. The well-examined program also addresses real-life issues like; e.g. "do parents have to be always present?" or, "what to do when parents are having different needs from those the baby does," and "when does pampering start, and which limits does an infant need, and when?" The program is for parents-to-be up to the seventh month of pregnancy, and it is continued after delivery until baby's first birthday; parents may continue up to the second or third birthday. There is a training of sensitivity toward the infant within a group in 10 days of class. Groups are run by two mentors in whole day seminars, 4 days during pregnancy, six after delivery. Stabilization and imagination exercises in stressful situations are conducted, especially in adaptation phase after delivery. A parental sensitivity training video supports the reading of signals and needs of the baby. A scientific foundation via attachment interviews with parents, diagnostic questionnaires, and other evaluation tools has recently led to first results [128].

4.1.3 "Skin-to-skin-care"/"kangaroo care"

An early experience of the infant's feedback is very important not only for intuitive parenting regulation but also for parental attachment behavior. The mother's feeling of self-efficacy evoked by the infant's feedback paves the way for relying on her intuitive competencies. One successful method to moderate early and unexpected separation of the infant from the mother's body, which can make both child and parents tend to insecure modes of bonding [129], is "skin-to-skin care," or "kangaroo care." Kangaroo care originally stems from the 1970s when Colombian mothers were advised to take their babies home and carry them on their chests for days and weeks. Through this intervention, infants were supplied with warmth and fed with milk [130]. Adapted to newborn intensive care unit (NICU) application, and incorporated in the NICU setting, "kangaroo care" became one of the most important care standards in developed countries nowadays [14, 131]. In the meantime, there have been many findings on the advantages of continuous bodily contact and on interaction between infant and parents. Recent findings on oxytocin and bonding add to a perspective of incorporating bodily and psychic factors; a recent study found lower depression scores in parents after giving neonatal massage [132]. It seems that people's ancient intuitive knowledge about bodily contact can be said to have been verified again and again; skin contact turned out to be highly important [58].

4.1.4 "NIDCAP®"

"NIDCAP®," i.e., Newborn Individualized Developmental Care and Assessment Program has been developed by Heidelise Als and her team members at Boston Children's Hospital. By distinguishing normal from abnormal neonatal behavior and in trying to obtain some prognostic conclusions about long-term development from newborn period behavior, Als became aware of the enormous influence that intensive care does have on the behavior of full-term and preterm newborn infants. Starting with these observations, the entire concept that should enable optimal development of each premature infant through individual care, and in spite of interfering intensive care treatment influences, was developed and patented [133, 134]. Neonatal care according to "NIDCAP®" principles has become more and more popular all over the world; it has been imported and implemented in Europe and is applied in the NICU at the Neonatology Unit at the University of Heidelberg, Germany. It has been designed for professionals that deal with preterm infants and their parents; its main issue is "reading the preterm infant" [135]. The individual intervention consists of daily (7 days a week) observation and evaluation of the infant's behavior, of support for care-givers in understanding the infant's stress and comfort signals, and of suggestions for parents and staff in terms of ways to support the infant's development, i.e., adjust their care according to these signals. The concept treats infants as active participants in the care provided, which is certainly most reasonable [14].

4.1.5 "Das Baby verstehen" (understanding your baby)

"Das Baby verstehen" [136] is a structured program for expectant mothers and their partners. Couples are supported through a midwife who will focus on the overall life situation of the family-to-be. Everyday communication between parents and their babies is illustrated in the instructions. The "reading" of the infant is at the center of most of the course lessons. Live video tapes support the instructions. Playful exercises will focus on the personal well-being of parents-to-be as well as on how to remain a couple when there will be three of them. In 2003 and 2004 the program was developed at the University of Heidelberg, followed by a revision in 2005, with accompanying evaluation in a German county district. The strengths and shortcomings of the expert trainings as well as of the courses for parents were explored, aiming at an integrative package of counseling for parents with infants up to the age of three. In this way, potential development of dysfunctional interaction in families is avoided early in order to prevent bodily and mental disorders in infants. The underlying concept has been depicted in a textbook of basic findings [137].

4.2 Approaching kindergarten age: "Faustlos"

Empathy as well as the competence to change one's perspective are key issues in the prevention of violent acting. In Germany, Mollenhauer et al. [54] elaborated on such in what can still be called a basic reference textbook on how socialization in family and society works, i.e., from a psychosocially integrated perspective. The second International Conference on Social-Emotional Learning, which took place at the University of Heidelberg [138] reactivated that perspective exploring both differences and similarities in countries and cultures, so that a multinational background makes sure concepts are compatible with each other [139].

A program for kindergartens like "Faustlos" [140–142], which has been designed for four-to-six-year-olds, seems to be most effective in preschoolers, yet even younger children participating in it will benefit as well. It is an adaptation of Seattle-based program, "Second Step" [143], translated to German-speaking countries as "no fists." The program has been developed and evaluated at the University of Heidelberg [144, 145]; a pre/post randomized control trial behavioral study proved the program to be effective especially as to a decrease of verbal aggression in children [15, 115, 116, 146, 147]. Competencies of self-regulation turned out to be of paramount importance, something, which is especially difficult in traumatized and insecurely attached children. Though not replacing therapy, "Faustlos" offers a wide variety of techniques and strategies for children to learn how to cope with inner impulses. Also, the program is conducted by constant relational persons in a closed group cycle of 1 year. This gives children a secure realm of learning and transfer in which no-one is excluded from the group. Instead, children learn from one another how to apply "Faustlos" competencies and dicta in everyday life. In order to increase favorable effects intergenerationally, the program makes use of involving parents reaching out to improve dealings with their children, regardless of age. Social-emotional learning aims at skills and competencies to be learned within an interactional framework. At the heart of "Faustlos" there are three issues to be transferred to children: getting to know empathy and the training to be empathic, learning to be capable of controlling one's impulses, and dealing with emotions of anger and rage. These issues are playfully dealt with by way of 28 continuous lessons. Each lesson contains a story that is told by the educator and is illustrated by an accompanying picture. Each lesson is structured the same way: at first, the topic of the lesson is outlined by playfully fantasying what the lesson will bring. Moreover, hand puppets (a toy dog and a toy snail) open up getting in contact with each other, further illustrating the issue of the lesson to come. This is followed by the actual lesson in which the story is told, is shown in the picture, and is discussed with the group. Role-playing, or alternative exercises at the end of the lesson will make sure the transfer to everyday life of children is initiated. Additionally, the educator is advised to return to the contents of the lesson during the following week. Ideally, one lesson per week is conducted. Since children learn how to cope with inner impulses, the range of possible reactions in stressful and conflict situations is broadened. Moreover, the aspect of mastering transitional stages seems quite important to both boys and girls participating in the program, which in the face of missing rites has a point in its own right [148, 149]. While male and female processes of individuation as reflected in ancient robinsonades show the male one to be rather abrupt and sometimes revolutionary, the female rather processual and preserving—still it is transformation proper [7]—in programs like this, transition as a developmental process should be examined.

5. Evaluation

Along the diachronic developmental perspective of the approaches depicted above, aside from "NIDCAP®," the "Faustlos" program for children has probably been evaluated the most, leading to augmentations in elementary and secondary education [141, 150, 151]. As to the kindergarten curriculum, identifying emotions turned out to be easier for children who took part in the program than for those who did not; the same for pro-social dealings with conflicts. The change of perspective through stories seen from different personal viewpoints is strongly supported in the program; something which has regularly been reported as revelatory [152, 153] as it calls attention to divergent experiencing. Generally, a specific anxiety-reducing effect supporting the transfer of competencies to everyday life has been shown in the program [154], which is highly important since effects on the level of intrapsychic emotion entail even more appropriate interpersonal pro-social behavior [91, 153]. Moreover, it has widely been well-accepted and therefore has been implemented at many kindergartens in Germany.

Practically, maternal symptom burden relief remains a highly important goal of intervention in the other approaches above. As has been shown, symptom relief has direct impact on the infant's development [120]. Personal reactivation and repetition of one's own experiences, such as preverbal, maybe even intrauterine [123] strain and other conflict formations leave their imprint on mothers-to-be: what can be said is that programs starting prenatally will approach mothers-to-be relatively early. This holds true for "SAFE" and "Das Baby verstehen," which are well-structured and designed for parents, and tend to address important everyday dealings with the infant such as the reading of signals in a closed or half-open group setting, with different emphases, respectively [128, 137]. Somewhat different from these, "Mutter-Kind-Bindungsanalyse" has been conceptualized as an individually shaped setting in which the emotionality of mothers-to-be and their empathic dealing with the unborn are approached in mid-pregnancy. Regarding this concept, the main case study results are promising [123, 155]. Especially combining of any of the structured programs with mother-infant bonding analysis would be worth studied. As there is much diversity in parents' perceptions of cause of children's symptoms [51], especially an early introspective psychosomatic intervention like mother-infant bonding analysis is promising. It has been recommended in particular by neonatology experts that have intensely applied "skin-to-skin/kangaroo care" or "NIDCAP®" [8]. The most important benefit from "skin-to-skin/kangaroo care," as studies have shown, is a change in the mother's perception of her child, attributable to the skin-to-skin contact ("bonding effect"), which supports and promotes attachment between infant and mother. Mothers in "skin-to-skin/kangaroo care" feel more competent ("resilience effect") in stressful situations in the NICU [14, 156–158], and mother-infant relations develop better; a tendency to less interaction disorders and less crying at the age of 6 months has been observed.

"NIDCAP®" has shown to have numerous positive effects on both the somatic and the neurological short-term development as well as on long-term developmental outcome of preterm infants such as motor and mental development, development of intelligence, behavioral development, and mother-infant interaction [159–162]. It also showed the first in vivo evidence of positive effects of early postnatal experience on brain development, i.e., of enhanced brain function and structure [11, 14]. This study demonstrated that the quality of the unborn's experiencing influences brain development significantly. Recently, further studies in the field have been conducted, such as on the effects of music and the mother's voice [14, 131, 163].

6. Conclusion

From perspectives of pre- and postnatal development, further research should be in what Panksepp termed affective neuroscience [164], an approach that does not deny drive and instinct and is most compatible with a psychoanalytic perspective. The findings of psychoanalysis and mental health research view affect as pivotal driving force in human development. That is why an ethological perspective will be helpful too, like in attachment research [165], or in the behavioral biology of Csikszentmihalyi showing that psychic satisfaction is in the process of pursuing, i.e., in anticipation itself [166]. Savoring the anticipation of something ahead is constitutive of the psychoanalytic process [167]; it is in itself psychotherapeutic, and it might be an effective factor in the programs depicted above. Still cognitive perspectives must be taken into account, on grounds of relational and phenomenological approaches [168] in connection with setting, context, and background [72].

There have been illuminating descriptions of the processes taking place in psychoanalysis [29, 81, 167]; many of these might analogously be examined in order to conceptualize objective processes of how subjectivity formation is affected by social objectivity [95]. It is not out of the question that the findings on mirror neurons can contribute to depicting such connection [36]. It should also be obvious that both biological factors and cultural upbringing have effect on the subject's development. The well-known problems of recent subjectivity formation have been documented culturally and clinically. We are dealing with the paradox of an inherent incompatibility in the "subiectum," in that it is underlying and at the same time subjugated; this means any absolutization will lead to aporia [100]. The question of scientific approaches, which at the time are dominated by relatively strict empiricist accesses and default interpretive accesses reveals limitations. A good balancing of quantitative and qualitative findings will be necessary to meet what can truly be called comprehensive psychology of human behavior, which lies in a combination of neuroscience and interpretation on grounds of reasonable concepts.

Author details

Götz Egloff[1*] and Dragana Djordjevic[2]

1 Practice for Psychoanalysis and Psychotherapy, Mannheim, Germany

2 University Children's Clinic, Niš, Serbia

*Address all correspondence to: g.egloff.medpsych.ma@email.de

References

[1] Kaplan-Solms K, Solms M. Clinical Studies in Neuro-Psychoanalysis. 2nd ed. New York/London: Karnac; 2002

[2] Solms M, Turnbull OH. What is neuropsychoanalysis? In: Weigel S, Scharbert G, editors. A Neuro-Psychoanalytical Dialogue for Bridging Freud and the Neurosciences. Cham/Heidelberg: Springer; 2016. pp. 13-30

[3] Stierlin H. Ich und die Anderen. In: Psychotherapie in einer sich wandelnden Gesellschaft [I and the Others]. Stuttgart: Klett-Cotta; 1994

[4] DeMause L. Was ist Psychohistorie? Eine Grundlegung [Foundations of Psychohistory]. Giessen: Psychosozial-Verlag Haland & Wirth; 2000

[5] Egloff G. Die psychogenetische Geschichtstheorie von Lloyd deMause [Lloyd deMause's Psychogenetic Theory of History]. In: Conference Presentation, Dept. Psychosomatische Kooperationsforschung und Familientherapie, Psychosomatische Klinik at Universitaetsklinikum Heidelberg. 2004

[6] Janus L. Transformations in emotional structures throughout history. The Journal of Psychohistory. 2016;**43**(3):187-199

[7] Janus L. Homo foetalis et sapiens. Das Wechselspiel des foetalen Erlebens mit den Primateninstinkten und dem Verstand als Wesenskern des Menschen [The Interplay of Fetal Experience with Primate Instincts and the Mind as Essence of Mankind]. Heidelberg: Mattes; 2018

[8] Linderkamp O. Gehirnentwicklung und fruehe Foerderung [Brain development and its early support]. In: Evertz K, Janus L, Linder R, editors. Lehrbuch der Praenatalen Psychologie [Textbook Prenatal Psychology]. Heidelberg: Mattes; 2014. pp. 19-33

[9] Roth G. Wirkung vorgeburtlicher und frueh-nachgeburtlicher Umwelteinfluesse auf Gehirn und Psyche [Effects of prenatal and early postnatal environmental influences on brain and psyche]. Psychotherapeut. 2016;**61**(2):118-123

[10] Fuchs T. Gibt es eine leibliche Persoenlichkeitsstruktur? Ein phaenomenologisch-psychodynamischer Ansatz [Is there a bodily personality structure? A phenomenological-psychodynamic approach]. In: Lang H, Faller H, Schowalter M, editors. Struktur—Persoenlichkeit—Persoenlichkeitsstoerung. Koenigshausen & Neumann: Wuerzburg; 2007. pp. 75-86

[11] Als H, Duffy FH, McAnulty GB, Rivkin MJ, Vajapeyam S, Mulkern RV, et al. Early experience alters brain function and structure. Pediatrics. 2004;**113**:846-857

[12] Freud S. Die Traumdeutung. GW. Vol. 2/3. Frankfurt: S. Fischer; 1900

[13] Papoušek H, Papoušek M. Intuitive parenting: A didactic to the infant's integrative competence. In: Osofsky JO, editor. Handbook of Infant Development. 2nd ed. New York: Wiley; 1987. pp. 669-720

[14] Djordjevic D. Premature Born Infant's Reaction to the Mother's Voice in Comparison to their Reaction to Music: Effects on Heart Rate and Heart Rate Variability. Heidelberg: University of Heidelberg; 2010

[15] Egloff G, Djordjevic D. On pre- and postnatal mental health intervention concepts. In: Williams J,

editor. Psychopathology—Symptoms, Challenges, and Current Concepts. New York: Nova Science; 2016. pp. 133-170

[16] Fernald A, Taeschner T, Dunn J, Papoušek M, Boysson-Bardies B, Fukui I. A cross-language study of prosodic modifications in mothers' and fathers' speech to preverbal infants. Journal of Child Language. 1989;**16**:477-501

[17] Spitz RA. Vom Saeugling zum Kleinkind [From infant to toddler]. Stuttgart: Klett-Cotta; 1967

[18] Spitz RA. Vom Dialog [On dialog]. Stuttgart: Klett Cotta; 1976

[19] Hellbruegge T. Zur Problematik der Saeuglings- und Kleinkinderfuersorge in Anstalten—Hospitalismus und Deprivation [On problems of infant and toddler care in institutions—hospitalism and deprivation]. In: Optiz H, Schmid F, editors. Handbuch der Kinderheilkunde [Textbook Pediatrics]. Berlin: Springer; 1966. pp. 384-404

[20] Hellbruegge T, Becker-Freyseng I, Menara D, Schamberger R. Deprivations-syndrome im Saeuglingsheim [Deprivation syndromes in infant institutions]. Münchener Medizinische Wochenschrift (1950). 1973;**115**:1753-1760

[21] Cierpka M, Windaus E, editors. Psychoanalytische Saeuglings-Kleinkind-Eltern-Psychotherapie, Konzepte—Leitlinien—Manual [Psychoanalytic Therapy of Infants, Toddlers, and Parents]. Frankfurt: Brandes & Apsel; 2007

[22] Juan S, Pescio N, Roussos AJ. Diagnostico Psicodinamico Operacionalizado 2 (OPD-2): Una revision bibliografica sobre su en investigacion empirica psicoanalitica. In: Memorias VI Congreso Internacional de Investigacion y Practica Profesional en Psicologia. University of Buenos Aires; 2014. pp. 187-188

[23] Sidor A, Koehler H, Cierpka M. Einfluss der soziooekonomischen Risikobelastung auf muetterliche Feinfuehligkeit, Stressbelastung und Familienfunktionalitaet [Impact of socioeconomic risk exposure on maternal sensitivity, stress and family functionality]. Praxis der Kinderpsychologie und Kinderpsychiatrie. 2018;**67**(3):257-273

[24] Janus L. Thoughts on some basic assumptions of psychoanalysis. In: Egloff G, editor. Child-Rearing: Practices, Attitudes and Cultural Differences. New York: Nova Science; 2017. pp. 197-220

[25] Lacan J. Schriften [Écrits]. Vol. 1. Olten/Freiburg: Walter; 1973 [first publ. 1966]. (Trans. G. Egloff.)

[26] Malabou C. Separation, death, the thing, Freud, Lacan, and the missed encounter. Avello Publishing Journal. 2012;**2**(1):1-36

[27] Freud S. Jenseits des Lustprinzips. GW. Vol. 13. Frankfurt: S. Fischer; 1920

[28] Freud S. In: Masson JM, editor. The Complete Letters of Sigmund Freud to Wilhelm Fliess, 1887-1904. Cambridge MA/London: Belknap/Harvard University Press; 1985

[29] Fink B. Introduction to Lacanian Psychoanalysis. Theory and Technique. Cambridge MA/London: Harvard University Press; 1997

[30] Sami-Ali M. L'espace imaginaire. Paris: Gallimard; 1974

[31] Janus L. Vom Kosmos zur Erde—vom Mythos zur Psychologie. Die Geschichte der Philosophie als Widerspiegelung der Evolution der Mentalitaeten und Lebensbezuege [From Cosmos to Earth—From Myth to

Psychology: The History of Philosophy as Reflection of Mentality Evolution]. Heidelberg: Mattes; 2019

[32] Egloff G. Ten theses on the spirit of terrorism. In: Morgan JB, editor. Focus on Terrorism. Vol. 14. New York: Nova Science; 2016. pp. 13-27

[33] Layton L. Some psychic effects of neoliberalism: Narcissism, disavowal, perversion. Psychoanalysis Culture & Society. 2014;**19**(2):161-178

[34] von Braun C. Fundamentalismus und Medien [Fundamentalism and the media]. Psychosozial. 2018;**41**(2):97-111

[35] Cosey FN. What "WALL-E" can teach us about global capitalism in the age of the anal father. The International Journal of Zizek Studies. 2018;**12**(1):1-19

[36] Rizzolati G, Fadiga L, Fogassi L, Gallese V. Resonance behaviors and mirror neurons. Archives Italiennes de Biologie. 1999;**137**:85-100

[37] Papoušek M. Persistent crying, parenting and infant mental health. In: Fitzgerald HE, Osofsky JO, editors. WAIMH Handbook of Infant Mental Health. New York: Wiley; 2000. pp. 415-453

[38] Reck C, Hunt A, Fuchs T, Weiss R, Noon A, Moehler E, et al. Interactive regulation of affect in postpartum depressed mothers and their infants: An overview. Psychopathology. 2004;**37**:272-280

[39] Ambrosini A, Stanghellini G. The concepts of "Depressive Pathogenetic Situation" and "Melancholic Type of Personality" as applied in the case of postpartum depression. In: Stoyanov D, editor. Psychopathology: Theory, Perspectives, and Future Approaches. New York: Nova Science; 2013. pp. 23-50

[40] Papoušek M, von Hofacker N. Disorders of excessive crying, feeding, and sleeping. The Munich interdisciplinary research and intervention program. The Infant Mental Health Journal. 1998;**19**:180-201

[41] Kuehner C. Psychiatrische Erkrankungen in Schwangerschaft und Stillzeit. Haeufigkeit, Verlauf und klinische Diagnostik [Mental disorders in pregnancy and postpartum. Prevalence, course and clinical diagnostics]. Nervenarzt. 2016;**87**(9):926-936

[42] Sidor A, Fischer C, Eickhorst A, Cierpka M. Influence of early regulatory problems in infants on their development at 12 months: A longitudinal study in a high-risk sample. Child and Adolescent Psychiatry and Mental Health. 2013;**7**(1):35

[43] Sidor A, Fischer C, Cierpka M. Early regulatory problems in infancy and psychopathological symptoms at 24 months: A longitudinal study in a high-risk sample. Journal of Child and Adolescent Behavior. 2013;**1**(3):10

[44] Zietlow AL, Eckstein M, Hernandez C, Nonnenmacher N, Reck C, Schaer M, et al. Dyadic coping and its underlying neuroendocrine mechanisms—Implications for stress regulation. Frontiers in Psychology. 2019;**9**(2600):1-10

[45] Papoušek H, Papoušek M. Biological basis of social interactions: Implications of research for an understanding of behavioral deviance. Journal of Child Psychology and Psychiatry. 1983;**24**:117-129

[46] Djordjevic D, Egloff G. Zur Bedeutung intuitiver Elternkompetenzen fuer die Entstehung von Bezogenheit [On the pertinence of intuitive parental competencies in creating human relatedness]. In: Langendorf U, Kurth W, Reiss HJ, Egloff G, editors. Wurzeln und Barrieren von Bezogenheit [Roots and Barriers of

Human Relatedness], Jahrbuch fuer Psychohistorische Forschung. Vol. 12. Heidelberg: Mattes; 2011. pp. 117-123

[47] Salecl R. The Tyranny of Choice. London: Profile; 2011

[48] Henningsen P, Martin A. Somatoforme Stoerungen [Somatoform disorders]. In: Herpertz SC, Caspar F, Mundt C, editors. Stoerungsorientierte Psychotherapie [Disorder-oriented Psychotherapy]. Urban & Fischer: Munich/Jena; 2008. pp. 541-559

[49] Goldbeck L, Bundschuh S. Subjektive Krankheitstheorien bei Kindern und Jugendlichen mit somatoformen Stoerungen oder Asthma bronchiale und ihren Eltern [Subjective illness beliefs of children and adolescents with somatoform disorder or asthma bronchiale and of their parents]. Praxis der Kinderpsychologie und Kinderpsychiatrie. 2007;**56**(1):3-18

[50] Ochs M, Seemann H, Franck G, Verres R, Schweitzer J. Familiaere Koerperkonzepte und Krankheitsattributionen bei primaeren Kopfschmerzen im Kindes- und Jugendalter [Body concept and cause attribution in families of children and adolescents with primary headache]. Praxis der Kinderpsychologie und Kinderpsychiatrie. 2002;**51**(3):209-223

[51] Shanley DC, Brown JB, Reid GJ, Paquette-Warren J. Parents' perceptions of their child's mental health problems. In: Barberis P, Petrakis S, editors. Parenting—Challenges, Practices, and Cultural Influences. New York: Nova Science; 2013. pp. 35-58

[52] Pinheiro-daCosta H, Vrabel JK, Zeigler-Hill V, Vonk J. DSM-5 pathological personality traits are associated with the ability to understand the emotional states of others. Journal of Research in Personality. 2018;**75**:1-11

[53] Schaefer A, Wimmer M, editors. Identifikation und Repraesentation. Wiesbaden: Springer VS; 1999

[54] Mollenhauer K, Brumlik M, Wudtke H. Die Familienerziehung [Family Socialization]. Munich: Juventa; 1978

[55] Villanova M. Conservatism, liberalism, and identity. The International Journal of Zizek Studies. 2017;**11**(1):1-12

[56] Fedor-Freybergh PG. Die Schwangerschaft als erste oekologische Situation des Menschen [Gestational period as first human ecological situation]. In: Janus L, Haibach S, editors. Seelisches Erleben vor und waehrend der Geburt [Psychic Experiencing Before and During Birth]. LinguaMed: Neu-Isenburg; 1997. pp. 15-21

[57] Fedor-Freybergh PG. Kontinuitaet und Dialog. In: Evertz K, Janus L, Linder R, editors. Lehrbuch der Praenatalen Psychologie. Heidelberg: Mattes; 2014. pp. 234-249

[58] Larguia AM, Gonzalez MA, Dinerstein NA, Conti CS. Perinatal programming prevention measures. In: Antonelli MC, editor. Perinatal Programming of Neurodevelopment. New York: Springer; 2015. pp. 425-441

[59] Haselbeck C. Foetale Programmierung: Der Einfluss von praenatalem Stress auf Entwicklung, Temperament und HHNR-Achsen-Funktion des Saeuglings—eine prospektive Studie [Fetal Programming: The Influence of Prenatal Stress on Overall Development, on Temperament, and on HPA Axis Function in Infants—A Prospective Study]. Kiel: University of Kiel; 2012

[60] Haselbeck C, Kulle A, Niederberger U, Bergmann TO, Steinmann E, Holterhus PM, et al. Foetale Programmierung: Der Einfluss von

praenatalem muetterlichen Stress auf Entwicklung und Temperament des Saeuglings [Fetal programming: The influence of prenatal maternal stress on development and temperament in infants]. Kindheit und Entwicklung. 2013;**22**(4):224-231

[61] Gluckman PD, Hanson MA. Developmental Origins of Health and Disease. Cambridge: Cambridge University Press; 2006

[62] Goebl C. Fetale Programmierung: Einfluesse der Fetalperiode auf die spaetere Entwicklung kardiometabolischer Erkrankungen [Fetal programming: Influences from the fetal period on cardiometabolic malfunction]. Speculum—Zeitschrift für Gynäkologie Geburtshilfe. 2015;**33**(3):10-12

[63] Huhn EA, Diemert A, Schleussner E, Hecher K, Arck PC. Fetale programmierung. In: Huppertz B, Schleussner E, editors. Die Plazenta. Grundlagen und klinische Bedeutung [Placenta: Basics and Clinics]. Berlin/Heidelberg: Springer; 2018. pp. 301-322

[64] Schwab M. Intrauterine Programmierung von Stoerungen der Hirnfunktion im spaeteren Leben [Intrauterine programming of adult brain dysfunction]. Gynäkologisch-Geburtshilfliche Rundschau. 2009;**49**(1):13-28

[65] Egloff G, Djordjevic D. Schwangerschaftshypertonie—eine psychosomatische Kasuistik [Hypertonia gravidarum: A psychosomatic case study]. Gyn. 2015;**20**(6):455-458

[66] Papoušek M, Schieche M, Wurmser H, editors. Regulationsstoerungen der fruehen Kindheit [Regulatory Disorders in Early Childhood]. Bern/Goettingen: Hans Huber; 2004

[67] Cierpka M, editor. Fruehe Kindheit: 0-3 Jahre [Early Childhood: Zero to Three]. Heidelberg: Springer; 2012

[68] Thiel-Bonney C. Fruehkindliche Regulationsstoerungen [Early infant regulatory disorders]. Monatsschrift Kinderheilkunde. 2009;**157**(6):580-586

[69] Bolten M. Diagnostik von Regulationsstoerungen im Saeuglings- und Kleinkindalter [Diagnostics of regulatory disorders in infants and toddlers]. In: Tschudin S, Maier B, Bitzer J, Ditz S, Rauchfuss M, editors. Grenzen wahrnehmen—respektieren—ueberwinden. Beitraege der 2, Drei-Laender-Tagung für Psychosomatische Gynaekologie und Geburtshilfe. Vol. 2012. Mabuse: Frankfurt; 2013. pp. 177-199

[70] Djordjevic D. Minimizing prenatal and perinatal psychic trauma in premature newborns—current interventions in Germany. In: Lecture at the 1st Intl. Congress on Psychic Trauma PTPPPA. Belgrade; 2015

[71] Egloff G. Integrative Psychoanalyse. In: Janus L, Kurth W, Reiss HJ, Egloff G, editors. Verantwortung fuer unsere Gefuehle [Responsibility for Our Emotions], Jahrbuch fuer psychohistorische Forschung. Vol. 16. Heidelberg: Mattes; 2015. pp. 303-331

[72] Welton D, Schueffel W. Toward a semantics of the symptom: The world of Frau D. In: Lohmar D, Brudzinska J, editors. Founding Psychoanalysis Phenomenologically. Heidelberg/New York: Springer; 2012. pp. 243-255

[73] Taubner S, Juen F. Gewalt in der Spaetadoleszenz. Perspektiven der Bindungsforschung [Violence in late adolescence: Perspectives of attachment research]. Psychotherapie und Sozialwissenschaft. 2010;**12**(2):59-77

[74] Gnazzo A, deCampora G, Khademi M, Guerriero V, Manoogian J, Kalayjian A, et al. The intergenerational impact of trauma: Individual, family and community implications. In: Williams J, editor. Psychopathology—Symptoms,

Challenges, and Current Concepts. New York: Nova Science; 2016. pp. 105-132

[75] Hurrelmann K. Aggression und Gewalt in der Schule—Ursachen, Erscheinungsformen und Gegenmassnahmen [Aggression and violence at school—Origins, phenomena, interventions]. Padagogisches Forum. 1992;**5**(2):65-74

[76] Brumlik M. Sigmund Freud (Vorwort) [Preface]. In: Freud S. Die Traumdeutung und andere Schriften [The Interpretation of Dreams, and Other Writings]. Zweitausendeins: Frankfurt; 2010. pp. 9-20

[77] Rochford L. Panic in childhood. In: Kilgus MD, Rea WS, editors. Essential Psychopathology Casebook. New York/ London: Norton; 2014. pp. 425-447

[78] Vance JE. Complex trauma. In: Kilgus MD, Rea WS, editors. Essential Psychopathology Casebook. New York/ London: Norton; 2014. pp. 448-474

[79] Schindler S. Entwicklung im Krieg. Empirische Daten zu Einfluessen auf die Kindheitsentwicklung [Development and war: Empirical data of influences on children's development]. In: Janus L, editor. Geboren im Krieg. Kindheitserfahrungen im 2. Weltkrieg und ihre Auswirkungen [Born in war: Childhood experiencing in World War II and its consequences]. Giessen: Psychosozial-Verlag Haland & Wirth; 2006. pp. 104-122

[80] Pena-Salazar C, Arrufat-Nebot FX, Alvarez MJ, Masramon-Fontsere H, Coll-Negre M, Roura-Poch P, et al. Do concepts of childhood trauma and dissociation provide an epistemological linkage between schizophrenia and PTSD? Nervenheilkunde. 2016;**35**:242-251

[81] Kessler H, Stasch M, Cierpka M. Operationalized psychodynamic diagnosis as an instrument to transfer psychodynamic constructs into neuroscience. Frontiers in Human Neuroscience. 2013;**7**(718):1-5

[82] Rank O. Die Analyse des Analytikers. In: Rank O, editor. Technik der Psychoanalyse. Vol. III. Giessen: Psychosozial-Verlag Haland & Wirth [first publ. 1931]; 2006. pp. 385-527

[83] Gondek HD. Subjekt, Sprache und Erkenntnis. Philosophische Zugaenge zur Lacanschen Psychoanalyse [Subject, language and knowledge: Philosophical approaches to Lacanian psychoanalysis]. In: Gondek HD, Hofmann R, Lohmann HM, editors. Jacques Lacan—Wege zu seinem Werk [Paths to his Oeuvre]. Stuttgart: Klett-Cotta; 2001. pp. 130-163

[84] Donnelly AM. Violence, power, and ideology. In: Donnelly AM, editor. Subverting Mainstream Narratives in the Reagan Era. New York: Palgrave Macmillan; 2018. pp. 1-12

[85] Cierpka M. Kinder mit aggressivem Verhalten. In: Ein Praxismanual fuer Schulen, Kindergaerten und Beratungsstellen [Aggressive Behavior in Children]. Goettingen: Hogrefe; 1999

[86] Levold T, Wedekind E, Georgi H. Gewalt in Familien. Systemdynamik und therapeutische Perspektiven [Violence in families: System dynamics and therapeutic perspectives]. Familiendynamik. 1993;**3**:287-311

[87] Graybow S, Eighmey J, Fader S. Privatization and psychoanalysis: The impact of neo-liberalism on Freud's tool of social justice. The Journal of Psychohistory. 2015;**42**(4):280-294

[88] Lacy HR. Lack of empathy. In: Kilgus MD, Rea WS, editors. Essential Psychopathology Casebook. New York/ London: Norton; 2014. pp. 579-603

[89] Taubner S. Gewalt als Symptom? Antisoziales Verhalten in der Adoleszenz

zwischen Normalitaet und Pathologie [Violence as a symptom? Anti-social behavior in adolescence between normalcy and pathology]. Psychosozial. 2008;**31**(3):101-117

[90] Taubner S, Curth C. Mentalization mediates the relation between early traumatic experiences and aggressive behavior in adolescence. Psihologija. 2013;**46**(2):177-192

[91] Morese R, Defedele M, Nervo J. I teach you to quarrel—Empathy and mediation: Tools for preventing bullying. In: Morese R, Palermo S, Nervo J, editors. Socialization—A Multidimensional Perspective. Rijeka: InTech; 2018. pp. 99-116

[92] Olivier B, Mos J. Biological Models of Aggression. In: Moeller HJ, van Praag HM, editors. Aggression and Autoaggression. Berlin/Heidelberg: Springer; 1992. pp. 10-33

[93] Žižek S. Violence: Six Sideways Reflections. London: Profile; 2008

[94] Žižek S. Absoluter Gegenstoss. Versuch einer Neubegruendung des dialektischen Materialismus [Absolute Recoil: Towards a New Foundation of Dialectical Materialism]. Frankfurt: S. Fischer; 2016

[95] McGowan T. Capitalism and Desire—The Psychic Cost of Free Markets. New York: Columbia University Press; 2016

[96] DeCampora G, Khademi M, Patel S, D'Onofrio E, Zavattini GC. The impact of socio-cultural factors on body image in two western countries. In: Conference Presentation, 43rd Annual Meeting of the Society for Cross-Cultural Research (SCCR). Charleston, SC; 2014

[97] Janus L. Die emotionale Dimension der Aufklaerung [The emotional dimension in enlightenment]. In: Janus L, Kurth W, Reiss HJ, Egloff G, editors. Verantwortung fuer unsere Gefuehle, Jahrbuch fuer psychohistorische Forschung. Vol. 16. Heidelberg: Mattes; 2015. pp. 13-47

[98] Egloff G. (Nichts) Neues vom unternehmerischen Selbst. Gegenrede zu M. Dornes, Macht der Kapitalismus depressiv? [(No) News from the entrepreneurial self: Objections to M. Dornes' Capitalism and Mental Illness]. *dap-hamburg.de*. 2017;**23**:1-11

[99] Maiso J. La subjetividad danada. Teoria Critica y Psicoanalisis [Damaged subjectivity]. Constelaciones—Revista de Teoria Critica. 2013;**5**:132-150

[100] Zima PV. Theorie des Subjekts. Subjektivitaet und Identitaet zwischen Moderne und Postmoderne [Theory of the Subject: Subjectivity and Identity between Modernity and Postmodernity]. 3rd ed. Tuebingen: Francke; 2010

[101] Resch F, Parzer P. Neue Morbiditaet und Zeitgeist [New morbidity and zeitgeist]. In: Braehler E, Herzog W, editors. Sozialpsychosomatik. Das vergessene Soziale in der Psychosomatischen Medizin. Stuttgart: Schattauer; 2018. pp. 319-334

[102] Rakonjac M, Dobrijevic L. The influence of early communication on child development and learning. In: Sovilj M, Subotic M, editors. Proceedings of the 4th International Conference on Fundamental and Applied Aspects of Speech and Language. Belgrade; 2013. pp. 368-371

[103] Papoušek M. Verwundbar, aber unbesiegbar—Ressourcen der fruehen Kommunikation in Eltern-Saeuglings-Beratung und -Psychotherapie [Vulnerable but invincible—resources of early communication in infant-parent counseling and psychotherapy]. In: Hellbruegge T, Schneeweiss B, editors. Fruehe Stoerungen behandeln—Elternkompetenz staerken [Treatment

of Early Disorders and the Support of Parental Competencies]. Stuttgart: Klett-Cotta; 2011. pp. 69-92

[104] Linder R. Psychosomatische Aspekte der drohenden Fruehgeburt [Psychosomatic aspects of imminent preterm delivery]. In: Janus L, Haibach S, editors. Seelisches Erleben vor und waehrend der Geburt [Psychic Experience Before and During Birth]. LinguaMed: Neu-Isenburg; 1997. pp. 73-78

[105] Linderkamp O. Das Fruehgeborene—der Fetus in der Intensivstation [The preterm infant: The fetus in the ICU]. In: Krens I, Krens H, editors. Grundlagen einer vorgeburtlichen Psychologie [Basics of Prenatal Psychology]. Goettingen: Vandenhoeck & Ruprecht; 2005. pp. 106-122

[106] Lorenz JM. Survival of the extremely preterm infant in North America in the 1990s. Clinical Perinatology. 2000;**27**:255-262

[107] Wolke D, Ratschinski G, Ohrt B, Riegel K. The cognitive outcome of very preterm infants may be poorer than often reported: An empirical investigation of how methodological issues make a big difference. European Journal of Pediatrics. 1994;**153**:906-915

[108] Wolke D, Meyer R. Cognitive status, language attainment, and prereading skills of 6-year-old very preterm children and their peers: The Bavarian Longitudinal Study. Developmental Medicine & Child Neurology. 1999;**41**:94-109

[109] Marlow N, Wolke D, Bracewell MA, Samara M. Neurologic and developmental disability at six years of age after extremely preterm birth. The New England Journal of Medicine. 2005;**352**:9-19

[110] Schlack HG. Neue Morbiditaet im Kindesalter: Aufgaben fuer die Sozialpaediatrie [New morbidity in children: Challenges for social pediatrics]. Kinderärztliche Praxis. 2004;**75**(5):292-298

[111] Erb L, Thiel-Bonney C, Cierpka M. Zum Zusammenhang von Fruehgeburtlichkeit und fruehkindlichen Fuetterstoerungen [Preterm delivery and early infant feeding disorders]. Ärztliche Psychotherapie und Psychosomatische Medizin. 2014;**9**(1):19-25

[112] Freud WE. Die Zweckmaessigkeit des Begriffs der Emotionellen Besetzung (Kathexis) fuer die Praenatale Psychologie [Expedience of the term cathexis for prenatal psychology]. In: Fedor-Freybergh PG, editor. Praenatale und Perinatale Psychologie und Medizin. Saphir: Stockholm; 1987. pp. 625-640

[113] Ziegenhain U, Kuenster AK. Fruehe elterliche Bindungsfoerderung in der interdisziplinaeren Zusammenarbeit [Support of early parental bonding in interdisciplinary cooperation contexts]. Familiendynamik. 2012;**37**(2):84-93

[114] Knapp C, Honikman S, Wirsching M, Husni-Pascha G, Haenselmann E. Adapting a perinatal empathic training method from South Africa to Germany. BMC Pilot and Feasibility Studies. 2018;**4**:101

[115] Egloff G, Djordjevic D, Cierpka M. Socio-emotional learning through Faustlos in Germany. In: Brekhman G, Sovilj M, Rakovic D, editors. Proceedings of the 1st Intl. Congress on Psychic Trauma PTPPPA. Belgrade; 2015. pp. 340-345

[116] Egloff G, Djordjevic D. Childhood intervention against violence. Facta Universitatis, Series: Medicine and Biology. 2017;**19**(1):1-5

[117] Seel NM, Hanke U. Erziehungswissenschaft [Textbook

Educational Science]. Heidelberg: Springer; 2015

[118] Benz M, Sidor A. Early intervention in Germany and in the USA: A comparison of supporting health services. An overview article. Mental Health & Prevention. 2013;**1**(1):44-50

[119] Taubner S, Wolter S, Rabung S. Effectiveness of early-intervention programs in German-speaking countries—A meta-analysis. Mental Health & Prevention. 2015;**3**(3):69-78

[120] Taubner S. Kompetenzentwicklung in den Fruehen Hilfen. Erkenntnisse aus der Psychotherapieforschung [Development of competencies in Fruehe Hilfen: Findings from psychotherapy research]. Psychotherapeut. 2015;**60**(5):446-450

[121] Jotzo M, Poets CF. Helping parents cope with the trauma of premature birth: An evaluation of a trauma-preventive psychological intervention. Pediatrics. 2005;**115**(4):915-919

[122] Hidas G, Raffai J. Nabelschnur der Seele [Umbilical cord of the soul]. Giessen: Psychosozial-Verlag Haland & Wirth; 2006

[123] Raffai J. Die psychoanalytische Arbeit mit Eltern noch ungeborener Kinder [Working psychoanalytically with parents of unborn children]. In: Poscheschnik G, Traxl B, editors. Handbuch Psychoanalytische Entwicklungswissenschaft. Psychosozial-Verlag Haland & Wirth: Giessen; 2016. pp. 397-409

[124] Klaus P. Kurzzeitpsychotherapie in der perinatalen Zeit zur Verringerung von psychischen und koerperlichen Symptomen und zur Erleichterung der Entwicklung der Eltern-Kind-Bindung [Perinatal short term psychotherapy for a decrease of psychic and bodily symptoms and for a support of parent-infant bonding development]. In: Brisch KH, Hellbruegge T, editors. Die Anfaenge der Eltern-Kind-Bindung [The Beginnings of Parent-Infant Bonding]. Stuttgart: Klett-Cotta; 2014. pp. 237-252

[125] Djordjevic D, Linderkamp O, Bruessau J, Cierpka M. Zusammenhaenge zwischen dem Wohlbefinden der Mutter und der Herzfrequenzvariabilitaet von Fruehgeborenen [Interrelation between well-being of the mother and heart rate variability of the preterm newborn]. Praxis der Kinderpsychologie und Kinderpsychiatrie. 2007;**56**:852-869

[126] Nelson K, Fivush R. The emergence of autobiographical memory: A social cultural developmental theory. Psychological Review. 2004;**111**(2):486-511

[127] Brisch KH. SAFE—Sichere Ausbildung fuer Eltern [Secure Education for Parents]. Stuttgart: Klett-Cotta; 2010

[128] Brisch KH, Landers S, Forstner B, Beck A, Quehenberger J. Bindung zwischen Eltern und Kind. Foerderung der Entwicklung einer sicheren Bindung mit dem SAFE-Programm [Attachment of parents and infant: Supporting the development of secure attachment through SAFE]. Padiatrie. 2016;**21**(3):29-34

[129] Klaus MH, Jerauld R, Kreger NC, McAlpine W, Steffa M, Kennell JH. Maternal attachment. Importance of the first post-partum days. The New England Journal of Medicine. 1972;**286**:460-463

[130] Doyle LW. Kangaroo mother care. Lancet. 1997;**350**:1721-1722

[131] Djordjevic D. Premature born infant's reaction to the mother's voice in comparison to their reaction to music. In: Fellow Lecture, Hanse Institute for Advanced Study (HWK). Delmenhorst; 2005

[132] Gnazzo A, Lella R, deCampora G, Guerriero V, Zavattini GC. The effect of neonatal massage on couples' perinatal depression symptoms. European Psychiatry. 2015;**30**(suppl. 1):1950

[133] Als H, Lester BM, Tronick EZ, Brazelton TB. Manual for the assessment of preterm infants' behavior (APIB). In: Fitzgerald HE, Lester BM, Yogman MW, editors. Theory and Research in Behavioral Pediatrics. Vol. 1. New York: Plenum Press; 1982. pp. 65-132

[134] Als H, McAnulty GB. The newborn individualized developmental care and assessment program (NIDCAP) with kangaroo mother care (KMC): Comprehensive care for preterm infants. Current Women's Health Reviews. 2011;7(3):288-301

[135] Als H. Reading the premature infant. In: Goldson E, editor. Nurturing the Premature Infant—Developmental Interventions in the Neonatal Intensive Care Nursery. New York: Oxford University Press; 1999. pp. 18-85

[136] Gregor A, Cierpka M. Das Baby verstehen [Understanding Your Baby]. Bensheim: Karl Kuebel; 2004

[137] Gregor A. Was unser Baby sagen will [What Our Baby is Trying to Say]. Munich: Ernst Reinhardt; 2007

[138] Cierpka M, Schick A, Ott I, Egloff G. 2nd International Conference on Social-Emotional Learning. Heidelberg: University of Heidelberg; 2001

[139] Egloff G. Report on the 2nd Intl. Conference on Social-Emotional Learning. Heidelberg: University of Heidelberg; 2001

[140] Cierpka M, Schick A, Ott I, Egloff G. Faustlos. Ein Curriculum zur Foerderung sozial-emotionaler Kompetenzen und zur Gewaltpraevention fuer den Kindergarten [Faustlos. A Curriculum for Kindergarten Supporting Social-Emotional Competencies and Preventing Violence]. Heidelberg: University of Heidelberg; 2002

[141] Schick A. Second step violence prevention and its European versions. In: Egloff G, editor. Child-Rearing: Practices, Attitudes and Cultural Differences. New York: Nova Science; 2017. pp. 99-114

[142] Berger U, Schick A. 16 Jahre Heidelberger Praeventionszentrum [16 Years of Heidelberg Prevention Center]. Psychotherapeut. 2018;**63**(5):355-360

[143] Beland K. Second Step: A Violence Prevention Curriculum. Seattle: Committee for Children; 1988

[144] Cierpka M, Egloff G, Schick A, Ott I. Pilotprojekt Faustlos Kindergarten. Heidelberg: University of Heidelberg; 2001

[145] Schick A, Ott I, Egloff G, Cierpka M. Faustlos Kindergarten. Ein Curriculum zur Gewaltpraevention an Kindergaerten. Kurzbeschreibung und Projektentwurf [Faustlos Kindergarten: A Curriculum for Deterring Violence. Project Description Draft Paper]. Heidelberg: Universitaetsklinikum Heidelberg; 2002

[146] Cierpka M, Schick A, Egloff G. Evaluation des Faustlos-Curriculums fuer den Kindergarten. Heidelberg: University of Heidelberg; 2004

[147] Cierpka M, Schick A. Die Evaluation des Faustlos-Curriculums fuer den Kindergarten. Praxis der Kinderpsychologie und Kinderpsychiatrie. 2006;**55**:459-474

[148] Egloff G. A "Snip" in time: Circumcision revisited. In: Olivier B, editor. Sexual Dysfunction. Rijeka: InTech; 2017. pp. 79-94

[149] Egloff G, Bender I, Roemer J. Conception and lifestyle—Integrative

psychosomatics in ovarian insufficiency. Journal of Andrology & Gynaecology. 2018;**6**(1):1-8

[150] Schick A, Cierpka M. Faustlos: Ein Gewaltpraeventions-Curriculum fuer Grundschulen und Kindergaerten [Faustlos: A curriculum for elementary school and kindergarten preventing violence]. In: Melzer W, Schwind HD, editors. Gewaltpraevention in der Schule [Violence prevention at school]. Nomos: Baden-Baden; 2004. pp. 54-66

[151] Schick A, Cierpka M. International evaluation studies of second step, a primary prevention programme: A review. Emotional and Behavioral Difficulties. 2013;**18**(3):241-247

[152] Egloff G, Cierpka M. Faustlos—Curriculum zur Gewaltpraevention. In: Conference Presentation, Congress "Adoleszenz—Bindung—Destruktivitaet" of the International Society for Adolescent Psychiatry (ISAP). Goettingen; 2002

[153] Cierpka M. Faustlos—wie Kinder Konflikte gewaltfrei loesen lernen [Faustlos: How Children Learn to Solve Conflicts Non-violently]. Freiburg: Herder; 2005

[154] Schick A, Cierpka M. Faustlos: Evaluation eines Curriculums zur Foerderung sozial-emotionaler Kompetenzen und zur Gewaltpraevention in der Grundschule. [Faustlos: Evaluation of a curriculum for social-emotional competencies and preventing violence for elementary school]. Kindheit und Entwicklung. 2003;**12**:100-110

[155] Schroth G. Die Bindungsanalyse nach Raffai und Hidas—eine Einfuehrung [Introduction to prenatal bonding analysis according to Raffai and Hidas]. International Journal of Prenatal and Perinatal Psychology and Medicine. 2009;**21**(3-4):343-347

[156] Fischer CB, Sontheimer D, Bauer J, Linderkamp O. Die Kaenguruh-Pflege Fruehgeborener. Stand der Forschung und Erfahrungen in Heidelberg [Kangaroo care of preterm infants: Heidelberg research findings and experiences]. Pädiatrische Praxis. 1997;**52**:609-618

[157] Fischer CB, Sontheimer D, Scheffer F, Bauer J, Linderkamp O. Cardiorespiratory stability of premature boys and girls during kangaroo care. Early Human Development. 1998;**52**:145-153

[158] Tessier R, Cristo M, Valez S, Giron M, Figueroa-de Calume Z, Ruiz-Palaez JG, et al. Kangaroo mother care and the bonding hypothesis. Pediatrics. 1998;**102**:17

[159] Als H, Lawhon G, Duffy FH, McAnulty GB, Gibes-Grossman R, Blickman JG. Individualized developmental care for the very-low-birth-weight preterm infant. Medical and neurofunctional effects. Journal of the American Medical Association. 1994;**272**:853-858

[160] Als H, Duffy FH, McAnulty GB. Effectiveness of individualized neurodevelopmental care in the newborn intensive care unit (NICU). Acta Paediatrica Scandinavica. Supplement. 1996;**416**:21-30

[161] Westrup B, Kleberg A, Wallin L, Lagercrantz H, Wikblad K, Stjernqvist K. Evaluation of the newborn individualized developmental care and assessment program (NIDCAP®) in a Swedish setting. Prenatal and Neonatal Medicine. 1997;**2**:366-375

[162] Westrup B. Newborn individualized developmental care and assessment program (NIDCAP)—Family-centered developmentally supportive care. Early Human Development. 2007;**83**:443-449

[163] Haslbeck F. Creative music therapy with premature infants: Testing a possible influence on brain structure, function, and development. Nordic Journal of Music Therapy. 2016;**25**(suppl. 1):32

[164] Panksepp J. Affective Neuroscience: The Foundations of Human and Animal Emotions. Oxford: Oxford University Press; 1998

[165] Ainsworth MDS, Blehar MC, Waters E, Wall S. Patterns of Attachment: A Psychological Study of the Strange Situation. Hillsdale: Lawrence Erlbaum; 1978

[166] Csikszentmihalyi M. Flow—The Psychology of Optimal Experience. New York: Harper Perennial; 2008

[167] Schmid M. Der psychoanalytische Akt. In: Gondek HD, Hofmann R, Lohmann HM, editors. Jacques Lacan—Wege zu seinem Werk. Stuttgart: Klett-Cotta; 2001. pp. 49-73

[168] Fuchs T. Das Gehirn—ein Beziehungsorgan [The Brain: A Relational Organ]. Stuttgart: Kohlhammer; 2008

Chapter 5

Mindsets and Failures: Neural Differences in Reactions to Mistakes among Second-Grade Finnish Girls

Ita Puusepp, Tuisku Tammi, Minna Huotilainen, Teija Kujala, Elina Kuusisto, Sonja Laine and Kirsi Tirri

Abstract

Mindsets have been identified as an important factor in explaining learning differences among students. Growth mindset students have been shown to recover from mistakes easier than fixed mindset students, and recent neuroscientific research has shown differences in the brain's event-related potentials to errors in fixed and growth mindset participants. The purpose of this study was to examine and evaluate these differences in the Finnish elementary school context. To achieve this, event-related potentials of ten 8-9-year-old female students, five of them with a fixed mindset and five with a growth mindset, were recorded during a go/no-go task. Differences between the two groups emerged; however, they were different from the results of some previous studies in the field. These findings are discussed in the light of earlier neuroscientific research related to mindsets, including limitations and suggestions for future research in the field.

Keywords: mindset, implicit belief, education, error monitoring, event-related potential, error-related negativity, error-related positivity, Finland, elementary school

1. Introduction

In this chapter, mindsets and differences in the neural mechanisms of attention allocation and other automatic reactions to errors between fixed and growth mindset students are discussed. The chapter presents results from a pilot study examining and evaluating these differences among girls in the Finnish elementary school context. These findings are discussed in the light of previous neuroscience research related to mindsets, including limitations of the studies conducted so far and suggestions for future research in this field.

Mindsets are implicit beliefs individuals hold about the malleability of basic qualities and abilities. People with a fixed mindset (the entity theory) believe human qualities are static; those with a growth mindset (the incremental theory) believe basic qualities can be developed [1]. The theory about mindsets helps us understand how people make sense of the world and their experiences [2].

IntechOpen

The theory can, for example, help us understand individual differences in goal pursuit, self-regulation, and response to feedback and setbacks by shedding light on how people construct meaning, interpret their experiences, and respond to their world. Indeed, there is a growing literature describing the connections between different mindsets to different behaviors and outcomes (e.g., see [3, 4]).

Mindsets are also highly relevant when it comes to the educational context. Indeed, in the last decades, they have been identified as an important factor in explaining learning differences among students [5]. Moreover, they seem to be especially relevant in certain academic domains, such as mathematics [6, 7]. Mathematics seems to be a subject about which people tend to hold more of a fixed mindset when compared to other educational subjects [6, 8]. Indeed, compared to achievement in social science and other subjects, achievement in mathematics is often believed to depend more on an innate ability that is uncontrollable [8]. Interestingly, holding a growth mindset about mathematical ability seems to be especially beneficial for girls when compared to boys, leading to higher grades in math [9]. Thus, as growth and fixed mindsets seem to be differentially related to the students' academic outcomes, the effort they put into learning, and the way students cope with setbacks and failures, it is highly important to consider and address mindsets in the educational context [7, 10–12].

In order to shed more light on mindsets and how they affect behavior, there has, in the recent years, been a growing interest in understanding the mechanisms behind the relations between mindsets and behavioral outcomes, including interest in the possible neural mechanisms that are involved in these processes [13–17]. Indeed, individuals with a growth mindset tend to recover from setbacks easier than individuals with a fixed mindset, and neural activity concerning automatic reactions to errors seems to be involved in this ability to rebound from mistakes (for review, see [18]). Although, thus far the neuroscientific research related to mindsets is still rather scarce, especially concerning studies conducted on children. We found only two studies connecting neuroscience and the theory of mindsets, which have focused on children [15, 17].

Most of the neuroscientific studies on mindsets have examined the connections between mindsets and electroencephalogram (EEG) recordings, more specifically the connections between mindsets and event-related potentials (ERPs) [13, 14, 16, 17]. Mangels and colleagues [13] had the participants of the study answer general knowledge questions and used EEG recordings to measure their neural responses to the feedback for the questions. In other studies [14, 16, 17], the researchers used a go/no-go or Flanker's task and measured the participants' neural responses to errors. All of these studies showed differences in the neural mechanisms, more specifically in the ERPs, of fixed and growth mindset participants, which might reflect differences in the processing of errors and feedback between fixed- and growth-minded participants. More specifically, researchers [14, 17] have found growth mindset to be related to an enhanced amplitude of the error-related positivity (Pe) component of ERPs, with no differences in the amplitude of error-related negativity (ERN). In study [13], growth and fixed mindset participants differentiated in the anterior frontal P3 to negative performance-relevant feedback, which might refer to negative feedback having a stronger affective effect in the case of a fixed mindset. In study [16] P3 amplitude was larger, and late Pe amplitude was smaller in participants with an induced growth mindset when compared to the participants with an induced fixed mindset. In addition to the studies using EEG recordings, there are two studies that have used functional magnetic resonance imaging (fMRI) to explore the neural mechanisms connected to mindsets [15, 19].

At the same time, even though these neural differences between growth and fixed mindset have been shown to be present among undergraduates and children

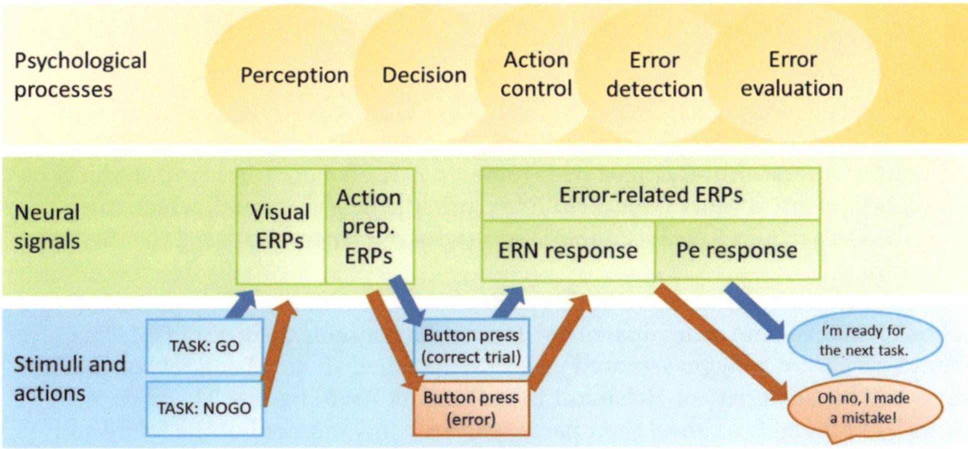

Figure 1.
Visual representation of the research design.

in North America, we found only one neuroscientific study on mindsets that has addressed different cultural contexts [19]. This study focused on mindsets about emotion regulation and not about intelligence. Still, results from that study and other previous raise questions about the cultural dependency and context of mindsets and their relations and, thus, point to the need for research on mindsets also in different cultural contexts [7, 20, 21]. This discussion illustrates the importance of investigating mindsets and their neural mechanisms also in different cultural contexts.

Taking into account the previous discussion and the stated importance of connecting psychological, educational, and neuroscientific research when studying mindsets [18], the purpose of our pilot study was to examine and evaluate the neural differences of attention allocation to mistakes between growth and fixed mindset girls in the Finnish elementary school context. Relying on the previous research in this field, we expected to detect differences in the error-monitoring ERPs of growth and fixed mindset participants. For this ERN and Pe were recorded. ERN has been associated with immediate, perhaps unconscious, error-correction or simply conflict-detection processes [22, 23]. Pe has been associated with conscious error awareness, attention allocation to errors [22], and conscious processing of motivationally significant events [24]. It has been suggested that Pe possibly reflects a subjective emotional error assessment process, which could be modulated by the individual significance of the error [23, 25]. As can be seen in **Figure 1**, at the psychological level, we assume that several processes take place, related to perceiving the task, making decision about the response, performing the action, detecting whether the action was right or wrong, and, finally, in the case of an error, evaluating the error and its consequences. At the level of the neural signals or ERPs, we can measure responses related to visual perception and action preparation (not reported in this study due to the averaging according to button press), the Pe response and the ERN response. These responses depend on the task (go trial or no-go trial), the action (button pressed or not pressed), and the correctness of the button press and are expected to also depend on the mindset of the participant.

2. Methods

Participants of the study were 10 right-handed second-grade female students aged 8–9 years (mean = 8.50, SD = 0.53). All of the participants were native Finnish speakers and students from a Finnish public elementary school, namely, the Viikki Teacher Training School of the University of Helsinki, where the student teachers practice under the guidance of mentors who are highly skilled in teaching. Additionally, research, practice, and development activities have a crucial role in Viikki Teacher Training School. The school has learning resources available for different learners with advanced pedagogies in use. Elementary school students in Viikki School are in general local children from the neighborhood, which can be described as a medium socioeconomic status district when compared to other areas in Helsinki [26].

The students' participation in this pilot study was voluntary, and parental, school principal, and municipal officials' written consents were obtained. The study was part of a bigger research project, which had already been reviewed and approved by University of Helsinki Ethical Review Board before. The participants had the right to cancel their participation at any moment of the study and measurements.

Participants had previously been classified as growth or fixed mindset students in the following manner: during individual interviews a researcher had asked the students 10 questions of a 5-point Likert-type scale questionnaire based on Gunderson and colleagues' mindset questionnaire used among children in previous research [27, 28]. They were also asked to describe how they understand the words "intelligence" and "giftedness." During that interview the participants were encouraged to bring up examples or questions related to the questionnaire.

The experiment was conducted by two experimenters during the school day in a separate space at the school premises. Before the experiment, the students were briefed about the process; they were encouraged to ask questions about the experiment and were reminded that they can cancel their participation at any moment. Participants then completed the task on a laptop. After the task, participants were debriefed about the experiment and compensated. The whole procedure lasted for approximately 1 h per participant.

The task was an age-appropriate go/no-go task adapted from Grammer and colleagues' study [29]. Participants were told that the task was a game in which they had to help a zookeeper catch animals and were instructed to press a button every time they saw a picture of an animal (go trial) except when the animal was an orangutan (no-go trial), because orangutans were also helping the zookeeper. The task consisted of a practice block (9 go trials, 3 no-go trials) followed by 16 blocks (30 go trials, 10 no-go trials) making up a total of 640 trials. Each stimulus was presented for 750 ms followed by a blank screen for 500 ms (response window 1250 ms). The participants were allowed small breaks between blocks and a longer one between blocks 8 and 9.

The task was conducted with presentation software (Neurobehavioral Systems, Inc., Albany, CA). EEG data were recorded with portable equipment (BrainVision QuickAmp amplifier) using 32 Ag-AgCl active electrodes (ActiCap, Brain Products, Germany) including two mastoid electrodes, one nose and one vertical eye movement electrode. Electrolyte gel (Signa Gel, Bio-Medical Instruments, Inc., Warren, MI) was used at each electrode. The data were recorded with BrainVision Recorder at 500 Hz sampling rate.

After recording, the EEG data were processed with Matlab R2017b software (Mathworks, Natick, MA) with EEGLAB 14.1.2b toolbox. The signal was high-pass filtered at 0.1 Hz and epoched 1250 ms before and 500 ms after response. In addition to visual inspection, artifactual epochs were rejected by detecting abnormal trends and abnormal spectra, and eye movement artifacts were removed using independent component analysis (ICA) [30]. The data were re-referenced to the average of the two mastoid electrodes. Response-locked grand average ERPs for channels Fz and Pz were calculated and baseline corrected by subtracting the mean amplitude from −150 to −50 ms pre-response. For figures, the waveforms were low-pass filtered using a Butterworth filter of order 3 with a cutoff frequency of 30 Hz.

Behavioral data from the go/no-go task included response accuracy and reaction time measures for each trial. These were further processed in R statistical software (version 3.4.3) and used to compute measures for post-error adjustments, following Moser and colleagues [14].

3. Results

The responses to correct trials and error trials differed in both groups. Moreover, as expected, differences of error-monitoring ERPs between growth and fixed mindset students emerged, suggesting different attention allocation to mistakes, which is believed to play an important role in bouncing back after failure (**Figure 2**). It can be seen from the data that the difference curve calculated between the correct and error trials was larger in children with fixed mindset when compared to children with growth mindset. In the frontal areas (observed at Fz channel) in the early latencies 100–200 ms after response (the button press), the ERN amplitude (calculated as the difference between positivity on error trials and relative to that on correct trials, see **Figure 3**) is larger in the children with fixed mindset. There is no difference in the shape or timing of the ERN response in the two groups.

The data also show clear differences between the groups in the Pe component, the difference signal calculated between the correct and the error trials in the parietal electrodes (observed at Pz channel) in later latencies (200–500 ms after response). Fixed mindset was associated with larger Pe difference than growth mindset.

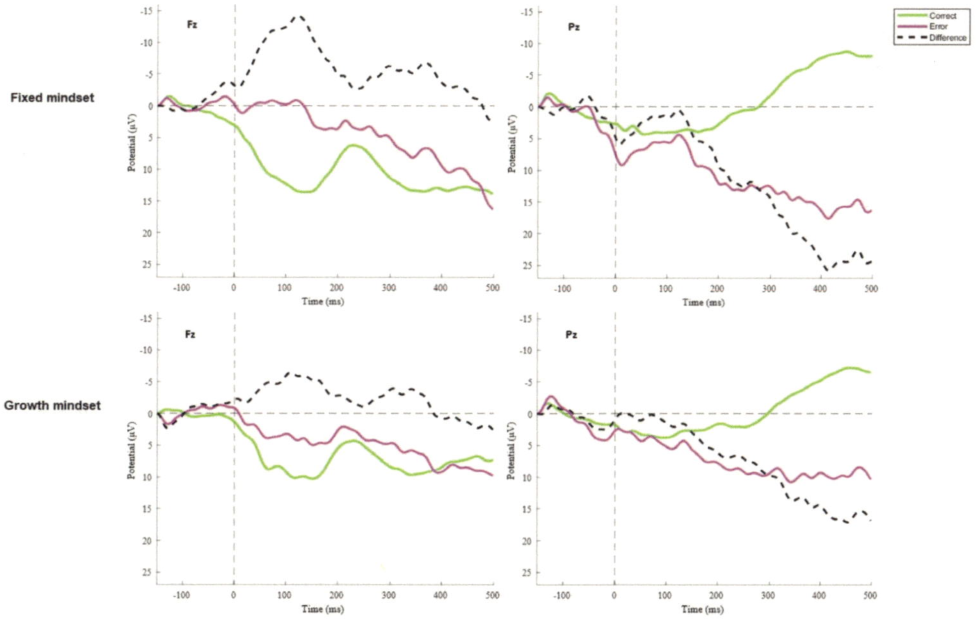

Figure 2.
Response-locked waveforms for correct and error trials in fixed (upper panel) and growth mindset groups (lower panel) at frontal Fz (left) and parietal Pz (right) electrodes.

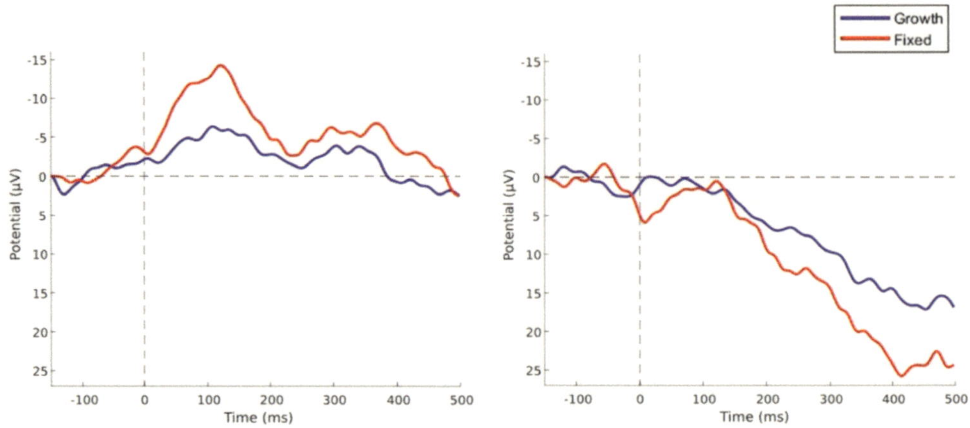

Figure 3.
Response-locked subtraction signals in fixed and growth mindset groups at frontal Fz (left) and parietal Pz (right) electrodes. Here, response to correct trials is subtracted from the response to the error trials.

At the behavioral level, growth mindset participants showed decreased post-error accuracy, meaning that they got less correct responses on trials following error hits than on trials following correct hits; this was opposite for the fixed mindset group. There was no considerable difference in post-error reaction times, but overall reaction times were shorter for the fixed mindset group, especially in error trials. Fixed mindset participants also made less error hits and more correct hits, i.e., their overall performance was slightly better. This is in line with results by Torpey et al. [31], who found that a more positive Pe is associated with greater accuracy and shorter reaction time in error trials. Overall, these results suggest that participants with a fixed mindset responded faster and, while allocating attention to errors, did not show improvement/adjustment in behavioral terms, such as post-error slowing.

4. Discussion

This pilot study contributes to the international mindset research by testing the mindset theory and experimental design, previously used in North America, in the Finnish context. It also provides evidence for differences in the neural mechanisms of attention allocation and in automatic reactions to errors between individuals with growth and fixed mindsets. Namely, in this study, the ERN amplitude was larger in the children with fixed mindset. Large ERN can be interpreted as more neural resources allocated to the detection of the error and also the further processing after detecting the error [32]. In addition to this, fixed mindset was also associated with larger Pe difference than growth mindset. These responses may reflect further processing of the errors, recovery after the errors, and reallocation of attentional resources to avoid future errors [33]. This suggests that fixed mindset children in this pilot study seem to invest a lot of effort in processing their errors and reorienting after the error has occurred. Growth mindset students also showed decreased post-error accuracy, while this was opposite for the fixed mindset group.

Interestingly, even though clear differences between the two groups emerged, these findings are somewhat inconsistent with the results from previously conducted research in North America [14, 17]. Namely, researchers [14, 17] have found growth mindset to be related to an enhanced amplitude of the Pe and better accuracy after mistakes, but not to ERN. Thus, the findings on the amplitude of Pe and also post-error accuracy were strikingly different from the findings from the North American studies. In addition to this, in this pilot study, differences in ERN were found, while this did not differentiate between growth and fixed mindset participants in the North American studies.

One possible explanation for this difference in the results of this pilot study, when compared to previous studies, is the young age of the participants. Namely, ERN seems to fluctuate during development [34]. Consistent with this, researchers [35] showed in their study that in younger children (8-to-10-year-old), a smaller ERN related to parent-reported anxiety, whereas in older children (11–13-year-olds), a larger ERN was significantly related to anxiety [35]. Consequently, the authors of the mentioned study discussed that it is possible that the relationship between increased error-related brain activity and anxiety may not emerge before early adolescence. Thus, one could speculate that it might be the same regarding the relationship between ERN and mindsets.

When discussing the differences between the results concerning Pe in this pilot study and previous studies, it is worth to mention that also Schroder and colleagues [17] showed that more attention allocation to errors (Pe) is not necessary for growth mindset children to recover from mistakes. Indeed, they did not find Pe to have the mediating role in recovering from mistakes as it had for grown-ups in the study conducted by Moser and colleagues [14]. Also the correlation found between growth mindset and Pe in study [17] on children was rather modest, and there were actually many growth mindset children who had average or below average Pe amplitudes. In addition to this, even though there is a difference in the time windows when compared to the current pilot study, in study [16] Schroder and colleagues found no differences in the early Pe (150–350 ms post-response time window) but found a smaller late Pe (350–750 ms post-response time window) amplitude in adult participants with an induced growth mindset when compared to the participants with an induced fixed mindset. Even though Pe has been shown not to have a similar age-related fluctuation as ERN [34], the inconsistencies of these findings might refer to other mechanisms involved in the processes of dealing with mistakes related to mindsets. Indeed, Meyer and colleagues also showed that smaller Pe amplitude related to greater parent-reported anxiety only among older children, with younger children's anxiety level having no significant effect on Pe [35]. Thus, taking into account the mentioned research concerning ERPs, it is possible to speculate that as the ERN fluctuates during development, a clearer relationship between increased error-related activity and mindset also may possibly not emerge before early adolescence, at least concerning ERN. The findings on Pe in this study, though, are somewhat controversial when compared to other studies and require further research on the developmental processes involved in error-related brain activity and mindsets, as the results suggest that there might be other mechanisms involved in the processes of dealing with mistakes when it comes to mindsets. Thus, in the future it would be important to conduct more research on the neural mechanisms related to mindsets among different age groups, including more participants and including both boys and girls as the current pilot study had a small sample size and only included girls as participants. Moreover, it would also be important to include participants from different schools and possibly more diverse socioeconomic backgrounds.

In addition to this, the results of this pilot study might differ from the previous ones due to a different cultural context. As mentioned in the first part of this chapter, there are studies that refer to possible culture- and context-dependency of mindsets [7, 19–21]. Thus, it would be important to study mindsets in different cultural contexts and also conduct comparative studies investigating mindsets, their functioning, and relations to neural mechanisms.

None of the neuroscientific research concerning mindsets has taken academic-domain-specificity into account. Previous studies using EEG recordings have measured mindsets about and used a task/test addressing general intelligence [13]; measured or induced mindsets about general intelligence [14, 16, 17] and the EEG measurements have been done during a completion of a go/no-go task or a Flanker's test. Even though the mindset measurement reflects the general underlying dimension of the mindset tendency in addition to the directly reflecting the mindset about intelligence [36], it is possible to speculate that the go/no-go task or Flanker's test used might not be reflecting the domain of intelligence for the participants. As these ERPs are measured and should theoretically reflect automatic reactions to errors of a person with a growth vs. fixed mindset, the ERPs may reflect the person's implicit beliefs in another domain than intelligence, which was measured or induced in these studies. Rather one could speculate that these tests might resemble more of a computer game than a test concerning intelligence, and thus, it might be more relevant comparing these ERPs regarding a growth vs. fixed mindset about the ability to play computer games, which might be remarkably different from the mindset that the individual holds about their intelligence or other domains like mathematics. Indeed, among these studies, as mentioned above, only Mangels and colleagues [13] have used a design, where the mindset measured and task used for EEG measurements match in their domains. Namely, they used measures of theories of intelligence (TOI) and a task, which included general knowledge questions. As mindsets, though, have been shown to have such considerable relations to academic outcomes [7], one important future direction would be measuring academic-domain-specific mindsets and using tasks/tests from the matching academic domain during the EEG measurements. This would enable to study the automatic reactions to errors in the specific academic domain of the held mindset and would thus yield to theoretically more sound results. One possibility to do this would be to modify the go/no-go task or Flanker's test to be more domain-specific, for example, resembling a math test and then comparing the ERPs from this test to the participants' academic-domain-specific (math-specific in the case of this example) mindsets.

All in all, understanding the neural mechanisms related to mindsets will enable, when combined with findings from other fields of research, the planning and construction of more successful interventions to encourage growth mindset. Taking into account the underlying neural mechanisms and structures of mindsets will enable to tap into how these implicit beliefs interact with cognitive and also other higher psychological processes, in order to improve students' learning experience and results. Moreover, it will help to understand how these interactions affect behavioral outcomes not only in the academic but also a variety of other contexts.

Notes

The earlier version of this chapter was presented in April 2019 as a talk at the International State-of-the-Art Symposium: Recent connections between Brain, Neuroscience and Education, which was part of the American Educational Research Association (AERA) Annual Meeting 2019 in Toronto, Canada.

Abbreviations

ERN	error-related negativity
ERP	event-related potential
Pe	error-related positivity

Author details

Ita Puusepp[1]*, Tuisku Tammi[2], Minna Huotilainen[3], Teija Kujala[4], Elina Kuusisto[5], Sonja Laine[6] and Kirsi Tirri[7]

1 Faculty of Educational Sciences, University of Helsinki, Helsinki, Finland

2 Cognitive Science, Department of Digital Humanities, University of Helsinki, Helsinki, Finland

3 Faculty of Educational Sciences and Cognitive Brain Research Unit, CICERO Learning Network, University of Helsinki, Helsinki, Finland

4 Cognitive Brain Research Unit, University of Helsinki, Helsinki, Finland

5 Department of Education, University of Humanistic Studies, Utrecht, The Netherlands

6 Faculty of Educational Sciences, Training Schools, University of Helsinki, Helsinki, Finland

7 Helsinki Collegium for Advanced Studies, University of Helsinki, Helsinki, Finland

*Address all correspondence to: ita.puusepp@helsinki.fi

References

[1] Dweck CS. Self-Theories: Their Role in Motivation, Personality, and Development. London, UK: Psychology Press; 2000

[2] Molden DC, Dweck CS. Finding "meaning" in psychology: A lay theories approach to self regulation, social perception, and social development. American Psychologist. 2006;**61**:192-203. DOI: 10.1037/0003-066X.61.3.192

[3] Sarrasin JB, Nenciovici L, Foisy L-MB, Allaire-Duquette G, Riopel M, Masson S. Effects of teaching the concept of neuroplasticity to induce a growth mindset on motivation, achievement, and brain activity: A meta-analysis. Trends in Neuroscience and Education. 2018;**12**: 22-31. DOI: 10.1016/j.tine.2018.07.003

[4] Schroder HS, Kneeland ET, Silverman AL, Beard C, Björgvinsson T. Beliefs about the malleability of anxiety and general emotions and their relation to treatment outcomes in acute psychiatric treatment. Cognitive Therapy and Research. Nov 2018:1-12. DOI: 10.1007/s10608-018-9985-7

[5] Kuusisto E, Laine S, Tirri K. How do school children and adolescents perceive the nature of talent development? A case study from Finland. Education Research International. 2017;**2017**:1-8. DOI: 10.1155/2017/4162957

[6] Gunderson EA, Hamdan N, Sorhagen NS, D'Esterre AP. Who needs innate ability to succeed in math and literacy? Academic-domain-specific theories of intelligence about peers versus adults. Developmental Psychology. 2017;**53**(6):1188-1205. DOI: 10.1037/dev0000282

[7] Costa A, Faria L. Implicit theories of intelligence and academic achievement: A meta-analytic review. Systematic Review. 2018;**9**:1-16. DOI: 10.3389/fpsyg.2018.00829

[8] Jonsson A-C, Beach D, Korp H, Erlandson P. Teachers' implicit theories of intelligence: Influences from different disciplines and scientific theories. European Journal of Teacher Education. 2012;**35**(4):387-400. DOI: 10.1080/02619768.2012.662636

[9] Degol JL, Wang M-T, Zhang Y, Allerton J. Do growth mindsets in math benefit females? Identifying pathways between gender, mindset, and motivation. Journal of Youth Adolescence. 2018;**47**:976-990. DOI: 10.1007//s10964-017-0739-8

[10] Aditomo A. Students' response to academic setback: "Growth mindset" as a buffer against demotivation. International Journal of Educational Psychology. 2015;**4**(2):198-222. DOI: 10.17583/ijep.2015.1482

[11] Blackwell LS, Trzesniewski KH, Dweck CS. Implicit theories of intelligence predict achievement across an adolescent transition: A longitudinal study and an intervention. Child Development. 2007;**78**(1):246-263. DOI: 10.1111/j.1467-8624.2007.00995.x

[12] Hong YY, Dweck CS, Chiu CY, Lin DMS, Wan W. Implicit theories, attributions, and coping: A meaning system approach. Journal of Personality and Social Psychology. 1999;**77**(3): 588-599. DOI: 10.1037/0022-3514.77.3.588

[13] Mangels JA, Butterfield B, Lamb J, Good C, Dweck CS. Why do beliefs about intelligence influence learning success? A social cognitive neuroscience model. Social Cognitive and Affective Neuroscience. 2006;**1**(2):75-86. DOI: 10.1093/scan/nsl013

[14] Moser JS, Schroder HS, Heeter C, Moran TP, Lee Y-H. Mind your errors: Evidence for a neural mechanism linking growth mind-set to adaptive posterror adjustments. Psychological

Science. 2011;**22**(12):1484-1489. DOI: 10.1177/0956797611419520

[15] Myers CA, Wang C, Black JM, Bugescu N, Hoeft F. The matter of motivation: Striatal resting-state connectivity is dissociable between grit and growth mindset. Social Cognitive and Affective Neuroscience. 2016;**11**(10):1521-1527. DOI: 10.1093/scan/nsw065

[16] Schroder HS, Moran TP, Donnellan MB, Moser JS. Mindset induction effects on cognitive control: A neurobehavioral investigation. Biological Psychology. 2014;**103**:27-37. DOI: 10.1016/j.biopsycho.2014.08.004

[17] Schroder HS, Fisher ME, Lin Y, Lo SL, Danovitch JH, Moser JS. Neural evidence for enhanced attention to mistakes among school-aged children with a growth mindset. Developmental Cognitive Neuroscience. 2017;**24**:42-50. DOI: 10.1016/j.dcn.2017.01.004

[18] Tirri K, Kujala T. Students' mindset for learning and their neural underpinnings. Psychology. 2016;7:1231-1239. DOI: 10.4236/psych.2016.79125

[19] Qu Y, Telzer EH. Cultural differences and similarities in beliefs, practices, and neural mechanisms of emotion regulation. Cultural Diversity and Ethnic Minority Psychology. 2017;**23**(1):36-44. DOI: 10.1037/cdp0000112

[20] Li Y, Bates TC. Does growth mindset improve children's IQ, educational attainment or response to setbacks? Active-control interventions and data on children's own mindsets. SocArXiv. Jan 2017;**23**:1-27. DOI: 10.31235/osf.io/tsdwy

[21] Zhang J, Kuusisto E, Tirri K. How teachers' and students' mindsets in learning have been studied: Research findings on mindset and academic achievement. Psychology. 2017;**8**:1363-1377. DOI: 10.4236/psych.2017.89089

[22] Nieuwenhuis S, Ridderinkhof KR, Blom J, Band GP, Kok A. Error-related brain potentials are differentially related to awareness of response errors: Evidence from an antisaccade task. Psychophysiology. 2001;**38**(5):752-760. DOI: 10.1111/1469-8986.3850752

[23] van Veen V, Carter CS. The timing of action-monitoring processes in the anterior cingulate cortex. Journal of Cognitive Neuroscience. 2002;**14**(4):593-602. DOI: 10.1162/08989290260045837

[24] Ridderinkhof KR, Ramautar JR, Wijnen JG. To PE or not to PE: A P3-like ERP component reflecting the processing of response errors. Psychophysiology. 2009;**46**:531-538. DOI: 10.1111/j.1469-8986.2009.00790.x

[25] Falkenstein M, Hoormann J, Christ S, Hohnsbein J. ERP components on reaction errors and their functional significance: A tutorial. Biological Psychology. 2000;**51**:87-107. DOI: 10.1016/S0301-0511(99)00031-9

[26] Vilkama K, Lönnqvist H, Väliniemi-Laurson J, Tuominen M. Differentiating metropolitan area: Socioeconomical differences between neighborhoods 2002-2012. 2014;**1**:53

[27] Gunderson EA, Gripshover SJ, Romero C, Dweck CS, Goldin-Meadow S, Levine SC. Parent praise to 1- to 3-year-olds predicts children's motivational frameworks 5 years later. Child Development. 2013;**84**:1526-1541. DOI: 10.1111/cdev.12064

[28] Gunderson EA, Sorhagen NS, Gripshover SJ, Dweck CS, Goldin-Meadow S, Levine SC. Parent praise to toddlers predicts fourth grade academic achievement via children's incremental mindsets. Developmental Psychology.

2018;**54**(3):397-409. DOI: 10.1037/dev0000444

[29] Grammer J, Carrasco M, Gehring W, Morrison F. Age-related changes in error processing in young children: A school-based investigation. Developmental Cognitive Neuroscience. 2014;**9C**: 93-105. DOI: 10.1016/j.dcn.2014.02.001

[30] Delorme A, Makeig S. EEGLAB: An open source toolbox for analysis of single-trial EEG dynamics including independent component analysis. Journal of Neuroscience Methods. 2004;**134**(1):9-21. DOI: 10.1016/j.jneumeth.2003.10.009

[31] Torpey DC, Hajcak G, Kim J, Kujawa A, Klein DN. Electrocortical and behavioral measures of response monitoring in young children during go/no-go task. Developmental Psychobiology. 2012;**54**:139-150. DOI: 10.1002/dev.20590

[32] Boksem MAS, Meijman TF, Lorist MM. Mental fatigue, motivation and action monitoring. Biological Psychology. 2006;**72**:123-132. DOI: 10.1016/j.biopsycho.2005.08.007

[33] Escera C, Yago E, Alho K. Electrical responses reveal the temporal dynamics of brain events during involuntary attention switching. European Journal of Neuroscience. 2001;**14**(5):877-883. DOI: 10.1046/j.0953-816x.2001.01707.x

[34] Davies PL, Segalowitz SJ, Gavin WJ. Development of response-monitoring ERPs in 7- to 25-year-olds. Developmental Neuropsychology. 2004;**25**(3):355-376. DOI: 10.1207/s15326942dn2503_6

[35] Meyer A, Weinberg A, Klein DN, Hajcak G. The development of the error-related negativity (ERN) and its relationship with anxiety: Evidence from 8 to 13 year-olds. Developmental Cognitive Neuroscience. 2012;**2**:152-161. DOI: 10.1016/j.dcn.2011.09.005

[36] Hughes JS. Support for the domain specificity of implicit beliefs about persons, intelligence, and morality. Personality and Individual Differences. 2015;**86**:195-203. DOI: 10.1016/j.paid.2015.05.042

Chapter 6

The Cerebellum and Autism: More than Motor Control

Marta Fernández, Teresa Sierra-Arregui and Olga Peñagarikano

Abstract

Autism spectrum disorder is a neurodevelopmental disorder characterized by deficits in social cognition at its core. Human and animal studies converge in the existence of a network of key brain structures involved in the perception, integration, and coding of social cues. These structures mainly involve areas traditionally associated with cognitive function, such as the prefrontal cortex; processing of emotions, such as the amygdala; and motivation and reward, such as the nucleus accumbens. The cerebellum, conventionally associated with motor functions, is lately being considered as a key structure within the social circuitry. Cerebellar neuroanatomical alterations are among the most replicated findings in postmortem brain samples of patients with autism. In addition, cerebellar defects have been proposed to affect the functioning of distal brain areas to which the cerebellum projects. In fact, animal studies support the inclusion of the cerebellum as part of the brain network regulating social cognition and provide a mechanistic tool to study its function within the social network. In this chapter, we review current evidence from human and animal studies, opening a new avenue for further research.

Keywords: autism, social behavior, neural circuit, cerebellum, dopamine, VTA, NAcc, animal model

1. Introduction

Autism spectrum disorder (ASD) represents a group of heterogeneous neurodevelopmental conditions characterized by deficits in social cognition, together with the presence of restricted and/or repetitive patterns of behaviors, activities, or interests [1]. Social cognition refers to those cognitive processes that allow individuals to successfully navigate the challenges of living in a social group. Thus, a functional social cognitive system involves the integration of several domains of behavior including attention, memory, emotion, and motivation to be able to understand identity, potential actions, social hierarchy, and emotional status of a conspecific and therefore guide the appropriate behavioral response [2]. In autism, deficits in social cognition processes are found at multiple levels, such as failure to initiate or respond to social interactions, lack of interest in social situations, abnormal social approach, difficulties expressing and understanding verbal and nonverbal communication (i.e., body language and facial expressions), and problems adjusting behavior to different social situations, among others [1]. Autism affects roughly 1 in 59 children, becoming one of the primary mental health issues

IntechOpen

worldwide [3]. In addition to the main symptoms, ASD is usually associated with other behavioral and/or neurological problems, such as hyperactivity, epilepsy, aggression, irritability, sleep problems, gastrointestinal symptoms, and sensory processing abnormalities [4].

Although currently accepted to be highly genetic (over 90% of the risk of developing ASD is due to genetic variation) [5], the etiology of ASD is complex, and its genetic architecture is diverse. Common allelic variation with small effect sizes is responsible for most cases, while rare but highly penetrant mutations that usually lead to other syndromes associated with autism are observed in about 20% of the cases [6]. In addition to genetic factors, exposition to some environmental factors during prenatal periods has also been associated with autism. Some of the most replicated are the intake of valproic acid, a drug used to treat epilepsy, during pregnancy and maternal infections. In all, the combination of interactions between genetic predisposition and environmental factors will determine the development of the disorder [7]. Given the clinical and etiological heterogeneity of ASD, the investigation of its pathophysiology has been challenging. From a research point of view, the study of "single gene" causes of autism, although rare in the population, has been proven to be useful to understand its pathophysiology and develop targeted treatments. In addition, animal models of monogenic causes of autism are easily generated and constitute a critical component of research. Research from both human and animal studies converge in a series of key brain structures and circuits involved in social cognition and their dysfunction in autism. Within these circuits, the cerebellum, traditionally associated with movement control, is becoming an important player in the social brain network. In this chapter, we will first start by describing the social brain circuitry traditionally thought to be affected in autism; we will then present evidence for the role of the cerebellum as a new player in the social circuitry and its role in the pathophysiology of ASD; finally, we will present data from animal models of monogenic causes of ASD in which a cerebellar pathology has been described such as Fragile X syndrome *(FMR1)*, tuberous sclerosis syndrome (*TSC1/TSC2)*, and Phelan-McDermid syndrome *(SHANK3)*, supporting the role of the cerebellum in social defects. In all, we believe the current evidence grants the need of further research of the cerebellar role in ASD pathophysiology.

2. Neural circuits involved in social cognition

In humans, brain regions implicated in social cognition have been identified mainly by lesion studies or by functional magnetic resonance imaging (fMRI) detecting differential activation in response to social versus nonsocial cues. Accordingly, a network of key brain structures involved in the perception, integration, and coding of social cues have been identified, receiving the denomination of "the social brain." These structures comprise brain areas traditionally involved in *cognitive processes*, such as frontal and temporal cortices; *motivation and reward*, such as the basal ganglia; and *processing of emotions*, such as the amygdala [8–10]. Proper function of the social brain should be considered in terms of the coordinated activity of the neural network involving these structures [11]. Accordingly, individuals with autism have been reported to present structural and/or functional alterations in these areas. Aberrant cortical organization is a pathological observation commonly seen in postmortem brain tissue of individuals with ASD [12]. Connectivity studies using fMRI indicate alterations in the PFC, with increased local and decreased long-range connectivity [13], which might be accounted for by the observed disorganized cortical structure. Structural MRI studies report increased amygdala size in children with autism, and the extent of amygdala enlargement is correlated with social deficits [14]. However, fMRI studies conversely report both hyperactivation [15] and hypoactivation [8] of the amygdala in response to social stimuli. These contrasting results have been suggested to indicate either a failure of the amygdala to engage upon social stimuli or, alternatively, an overreaction to such stimuli, coding them as threatening. Both cases would in turn result in social withdrawal. Similar contrasting results have been found when studying the reward system in ASD, as some studies observe a generalized decreased activation of the NAcc independent of stimuli [16], and others indicate that the hypoactivation is stimulus-dependent as individuals with ASD react to cues that are salient to them but not necessarily to neurotypical individuals [17], indicating that in autism a different set of cues might be coded as salient.

The cerebellum, conventionally associated with motor functions, is lately being considered a key structure within the social circuitry [18]. In fact, cerebellar neuroanatomical alterations, including the reduced size and number of Purkinje cells, are among the most replicated findings in postmortem brain samples of individuals with autism [19]. In addition, cerebellar defects have been proposed to affect the functioning of distal brain areas to which the cerebellum projects [20]. For example, the PFC long-range connection deficits observed in ASD mentioned above include the cerebellum [13]. The *developmental disconnection hypothesis* of autism suggests that certain areas of the brain that normally connect to the frontal lobe become disconnected during development. Thus, a change in connectivity in a certain area could affect the functioning of other brain regions either through compensation or adaptation of nearby circuitries [21]. Along these lines, deficits in connectivity of the cerebellum could account for dysfunction in connected areas, being possible to affect some of the social brain structures previously mentioned.

Although human neuroanatomical and functional studies have been very useful in the identification of the brain regions involved in social cognition, animal studies are critical to understand how information is processed at the circuit and molecular level, from the perception of the stimulus to the expression of a behavioral response. The mouse (*Mus musculus*) is currently the most widely studied, mainly for practical reasons and technical amenability. In addition, as mammals, mice present the same key brain structures and express a wide range of social behaviors that can be easily measured in the lab [22]. A schematic representation of the brain structures and circuits implicated in social cognition processes in rodents is presented in **Figure 1**. Animal studies have corroborated the role of previously described structures in social behavior and have given insight into circuit function. For example, disruption of the ratio between cortical excitation and inhibition (E/I) has been extensively hypothesized to be a causal mechanism in autism [23]. Again, one must consider that the cortical E/I balance is a complex process controlled both locally and distally by neuromodulation from connecting circuits, arguing against the specificity of a certain structure as preferentially involved in ASD since alterations in distally connected regions could also lead to an altered cortical E/I ratio. Nevertheless, in 2011, the application of the recently developed optogenetic techniques allowed to test for the first time this hypothesis in vivo [24]. The authors found that increasing, but not reducing, the E/I balance in the PFC would lead to social dysfunction in mice. Similarly, recent studies have indicated the role of the reward system and, specifically, dopaminergic projections from the ventral tegmental area to the NAcc, in the modulation of social interaction. Dopamine (DA) is a neurotransmitter traditionally linked to movement control and reward processing, including social reward and domains of behavior that are modulated by two separated DA pathways. DA-producing neurons are located in two main brain areas, the ventral tegmental area (VTA) and the substantia nigra (SN). The *nigrostriatal pathway* originates in the

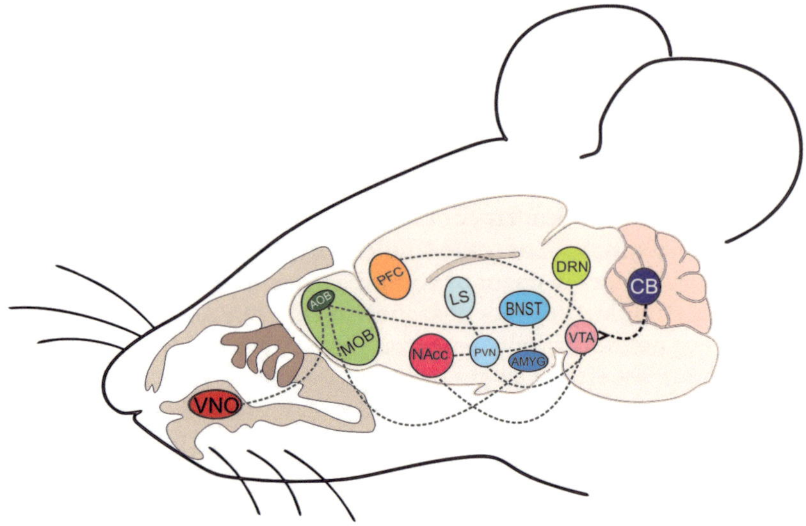

Figure 1.
Brain circuits linked to social cognition in mice. Olfactory signals from a social stimulus are perceived through the olfactory bulb and transferred to the amygdala (AMYG) and probably other structures to be processed. This social cue will modulate the activity of several structures. One of such, the ventral tegmental area (VTA), mainly composed of dopaminergic neurons, projects to the PFC (executive function) and NAcc (reward system). Of note, the cerebellum modulates VTA activity implying a role in social behavior and reward. Adapted from [28].

SN and projects to the striatum, modulating movement; and the *mesocorticolimbic pathway*, involved in cognitive processes including social cognition, originates in the VTA and is further subdivided into two pathways: the *mesocortical pathway*, which projects to the cortex, and the *mesolimbic pathway* which projects to limbic areas such as the NAcc [25]. Recently, it has been found that in mice, optogenetic stimulation of dopaminergic VTA-NAcc projections increased, while inhibition decreased during the time the animals were engaged in social exploration [26]. Interestingly, very recently Carta et al. [27] have demonstrated the role of the cerebellum in modulation of the reward pathway through direct control of the activity of the VTA, which could have profound implications for social behavior. The authors found that the cerebellum sends excitatory projections to the VTA and that optogenetic modulation of the cerebellum-VTA pathway bidirectionally modulated social behavior and reward. Considering previous studies where stimulation of VTA-NAcc DA projections modulates social behavior, it is highly likely that the cerebellum indirectly controls the activity of these projections. A cartoon deciphering the main brain structures and connections involved in social cognition in mice is presented in **Figure 1**.

3. The cerebellum

The cerebellum (Latin, *little brain*) is located in the posterior cranial fossa. Classically, the cerebellum has been linked to motor behaviors; however, more recent studies provide evidence for a role of the cerebellum in higher functions such as cognition, language, and social and affective behaviors [18].

3.1 Cerebellar anatomy

Structurally, the cerebellum is constituted of ten lobules: lobules I through V (which form the anterior lobe), lobules VI through IX (posterior lobe), and lobule X (flocculonodular lobe). Lobules VII and VIII are further subdivided (VIIA and VIIB and VIIIA and VIIIB); besides, the hemispheric extension of lobe VIIA is expanded and forms two major lobules, Crus I and Crus II. Dividing the cerebellum ventrally into two hemispheres, there is a central midline called the vermis (**Figure 2**) [20]. Within this anatomical division, a functional division according to the connection of each lobe can be found, establishing a topographic organization (see Subsection 3.2).

At the cellular level, the cerebellum is composed of an outer cortex constituted of gray matter, the cerebellar cortex, and an inner core formed mainly by white matter which encloses the deep cerebellar nuclei, the sole output channel of the cerebellum (see below). The cerebellar cortex is structured in three different cell layers: (1) *Molecular layer*: it is the outer layer, and it is composed of two types of cells, basket, and stellate neurons. Both are inhibitory and are part of the regulatory system of Purkinje cells. (2) *Purkinje cell layer*: it is located below the molecular layer. The Purkinje cells (PCs) represent the only output from the cerebellar cortex and are inhibitory. These neurons have a wide dendritic arbor, which extends to the top of the surface of the molecular layer, and project their axons to the deep cerebellar neurons in the inner cerebellar core, which are the only output of the cerebellum. (3) *Granular layer*: it is the deepest layer of the cerebellar cortex, and it is composed of excitatory granule cells which send their axons toward the molecular layer, forming the parallel fibers and making contacts with dendrites of PC. Granule neurons together with basket and stellate cells in the molecular layer constitute the main regulatory system of PC. Inhibitory interneurons—Golgi cells—are also found within the granular layer and act by altering the mossy fiber-granule cell synapse (see below). The Golgi cells receive input from the parallel fibers and provide an inhibitory feedback to the cells of origin of the parallel fibers (the granule cells).

Neurons in the deep cerebellar nuclei represent virtually all the output from the cerebellum. They receive inhibitory information from the PC and excitatory inputs from outside the cerebellum through the mossy and climbing fibers. The *climbing fibers are* originated in the brain stem (posterior part of the brain, continuous to the spinal cord, composed by the midbrain, the pons, and the medulla oblongata), particularly in the inferior olivary nucleus of the medulla oblongata. These axons

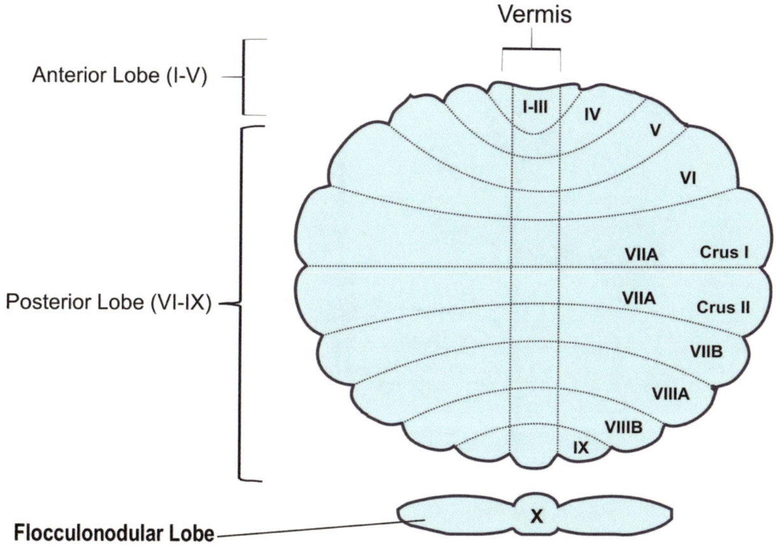

Figure 2.
Structural anatomy of the human cerebellum. Representation of the cerebellar lobules: anterior, posterior, and flocculonodular and their subdivisions. Lobules VII and VIII are further subdivided into lobules A and B. Crus I/Crus II constitutes further subdivisions of lobule VIIA. The vermis represents a midline separating the cerebellum in two hemispheres. Modified from [20].

make excitatory synapses with PC dendrites and with neurons in the deep cerebellar nuclei. The *mossy fibers* originate from several parts of the brain and spinal cord. These fibers form excitatory synapses with granule cells and with neurons of deep cerebellar nuclei. In turn, PCs receive two types of excitatory input from outside the cerebellum, one directly from the climbing fibers and the other indirectly via the parallel fibers of the granule cells (**Figure 3**) [29]. Anatomical investigations in animals and postmortem humans have established that cerebro-cerebellar connections are contralateral to each other and include an efferent cerebello-cortical pathway from the cerebellar nuclei to the cerebral cortex through the thalamus and an afferent cortico-cerebellar pathway through the pons.

3.2 Cerebellum connections and topography

The cerebellum has a unique topographic organization such as each region is attributed with a separate function based on their specific connectivity. Thus, the anterior lobe and lobule VIII contain the representation of the sensorimotor cerebellum; lobules VI and VII (including Crus I/Crus II and lobule VIIB) of the posterior lobe comprise the cognitive cerebellum; and the posterior vermis encompasses the limbic cerebellum. Dysfunction in the connection of these cerebellar areas with the spinal cord or cerebral regions will result in alterations in movement or cognitive functions, respectively [30, 31]. More specifically, the cerebellum has been proposed to have an important role in language by means of its connections with cortical areas implicated in this process. Studies using viral tracing in nonhuman primates report

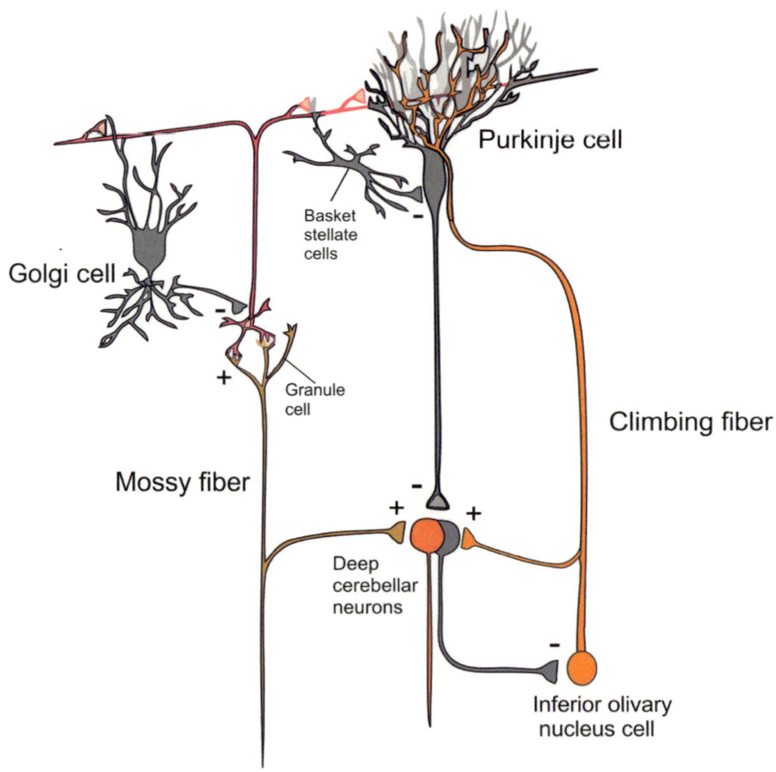

Figure 3.
Main cerebellar circuits. The mossy and climbing fibers carry the input information toward the cerebellum. The PC transmit the information to the deep cerebellar nuclei, which are the cerebellar output. The interactions between the cells are represented with (+) in case of excitatory connections and (−) when connections are inhibitory. Modified from [29].

the existence of strong connections between right Crus I/Crus II and different cortical regions implicated in language, such as Brodmann's area BA46 in the PFC [32]. In fact, dysfunction in cerebellar-prefrontal loops might underlie poorer performance on measures of language-related executive function in human patients with cerebellar abnormalities [33]. The cerebellum also seems to play a role in several social and affective processes. For example, imitation is a critical skill for implicit learning of social rules, and fMRI studies in humans show that during a task that involves observation and imitation of an action performed by a human model, activation of the Crus I/Crus II regions in the posterior cerebellum is increased [34]. Also, during a passive viewing paradigm, a stronger activation in the posterior cerebellum (lobules VI, VII, and X) has been found when comparing social versus nonsocial stimuli [35]. In fact, a recent meta-analysis of over 350 fMRI studies exploring the role of the cerebellum in social cognition supports that it plays a crucial role in several social paradigms such as mirroring (i.e., observation of human motion) and mentalizing (i.e., interpreting other people's thoughts and intentions) [36, 37].

4. Evidence that the cerebellum is involved in ASD

In the recent years, there has been increasing evidence showing a crucial role for the cerebellum in the etiology of ASD [38–40]. Although the field of cerebellar research in disorders of social cognition such as autism is still in its early stages, below we will describe the main structural and functional cerebellar abnormalities that have been described to date in autism, which provide strong evidence to grant further research.

4.1 Structural cerebellar abnormalities found in autism

The cerebellum is actually one of the most consistent sites of neural abnormalities found in autism [41]. Specifically, the reduced size and number of PCs are among the most replicated findings in postmortem brain tissue of individuals with autism [19]. This reduction in PC in autism patients has been found to be more pronounced in the Crus I/Crus II region of lobule VIIA [42]. Accordingly, a reduction in gray matter volume, smaller ratio of gray to white matter, and smaller vermis lobules VI–VII have been found in children with autism compared to controls [43, 44]. Further, in ASD patients, the degree of reduction in gray matter of Crus I/Crus II has been repeatedly found to correlate with the severity of symptoms in the social interaction and communication behavioral domains of ASD [39, 45]. Of note, some reports using adult brains indicate the presence of gliosis as an accompanying factor to the reduction of PC [46]. Other observed cerebellar cellular abnormalities are the presence of neuro-inflammatory processes [47].

Besides structural and cellular alterations, molecular abnormalities have also been reported in the cerebellum of ASD individuals. Alterations in the distribution of the mRNA levels of glutamic acid decarboxylase 67 (GAD67), an enzyme involved in the synthesis of the inhibitory neurotransmitter GABA, have been found. Thus, decreased GAD67 mRNA has been reported in PC [48], while increased GAD67 mRNA has been reported in cerebellar interneurons [49]. These cerebellar imbalances could account for the proposed E/I disequilibrium in ASD, as they could affect cerebro-cerebellar circuits [50].

4.2 Abnormalities of cerebellar function

Studies in humans using resting-state functional connectivity (rsFC) techniques have reported connectivity alterations between cerebellar and cortical areas in autism compared to typically developing individuals. Overall, a general cerebro-cerebellar over-connectivity has been found in the ASD group. However, both hyper-connectivity and hypo-connectivity have been reported depending on the regions analyzed. For example, cerebellar-sensorimotor FC (premotor and primary motor cortices, somatosensory temporal cortex, and occipital lobe) has been found to be atypically increased in ASD, while cerebellar-supramodal FC (prefrontal cortex, posterior parietal cortex, and inferior and middle temporal gyri) has been found to be decreased [51]. Analysis of cerebellar FC with language-related areas revealed a significantly reduced FC in ASD between the cerebellum and Broca's area and Wernicke's area [52], suggesting that the cerebellum plays a role in language functioning. Studies aiming at assessing the developmental pattern of cerebello-cerebral FC also report developmental alterations in ASD. FC between the cerebellum and subcortical regions was found to decrease in neurotypical individuals, while it increased in ASD [53]. It must be noted that no specific correlation between FC patterns and autism behavior has been detected, although reduced connectivity seems to be accompanied by an increase in the severity of the disorder [52], as assessed by the Social Communication Questionnaire, an ASD screening measure consisting of a brief (40-item) parent report that focuses on ASD symptomatology likely to be observed by a primary caregiver.

Few studies have investigated to date the FC between the cerebellum and cortical areas during task performance in ASD. During a sequential finger-tapping task, activations in motor circuits were found in both cases and controls. However, children with typical development showed activation of cerebellar structures that were silent in autistic children (lobules IV/V and anterior cerebellum). In addition, a reduced FC between premotor areas and the cerebellum was observed in autistic children, suggesting alterations in long-range cerebro-cerebellar connections [54]. In a task that requires perception and imitation of human actions, fMRI detected an engagement between the posterior superior temporal sulcus (pSTS) and the cerebellar region Crus I [55]. Interestingly, the degree of functional coactivation of pSTS and Crus I could predict social deficits in ASD in the "mentalizing skills" questionnaire, a parent report for specific social cognition skills based on imaginative mental activity that allows an understanding of the behavior of other people (intentions, needs, desires, or goals). Thus, stronger Crus I-pSTS interactions were associated with better mentalizing ability [55]. On a similar note, during a task that involves decoding the interactions between animated figures, aimed at examining the "theory of mind" network, that is, the ability to attribute mental states to others, a reduced cerebellar activation, particularly in Crus I, in participants with ASD was found [56]. Although many more studies are needed, overall the above-presented data indicate a role for cerebellar connections with key cortical social brain sites and, specifically for region Crus I/Crus II, in the pathogenesis of ASD.

5. Evidence from monogenic forms of ASD

The clinical and genetic heterogeneity present in ASD has made the study of the pathophysiology of the disease challenging. The study of genetically defined autism, as in the case of monogenic forms of ASD, which show a relatively homogeneous and well-characterized clinical manifestation, allows us to understand cellular and molecular mechanisms relevant to the disease. Although monogenic causes of ASD are often syndromic and not all patients with the syndrome show autistic features, by studying patients with and without ASD, we can start deciphering the pathomechanisms that lead to the disorder. As shown below, human postmortem and brain imaging studies of syndromic forms of ASD support the role of the cerebellum in the pathophysiology of the disease. Animal models of monogenic autism are easily generated and provide opportunities for direct manipulation of these brain regions and circuits to test their precise functions in social behavior paradigms. We will describe below the main syndromes with reported cerebellar dysfunction and supporting data from animal models (**Table 1**). In all, human and animal data point out a role for the cerebellum in social deficits in ASD.

5.1 Fragile X syndrome (FMR1 gene)

Fragile X syndrome (FXS) is the most common form of inherited intellectual disability, and ~30% of FXS patients are also diagnosed with ASD. FXS is considered also as the most common genetic cause of ASD, representing around 3% of ASD individuals. FXS arises from loss of function mutations in the X-linked FMR1 gene, which result in either total absence or functional inactivation of the encoded protein (FMRP), an mRNA-binding protein involved in translational regulation [65]. Although FXS affects many brain regions, cerebellar PC loss and cell displacement, as seen in idiopathic autism, have been reported in human postmortem studies of FXS [66]. Interestingly, imaging studies have identified specific abnormalities in the posterior cerebellar vermis (lobules VI–VII) in FXS patients with ASD that are not seen in FXS patients without comorbid ASD diagnosis. Further, these

Mouse model	Structural abnormalities	Functional abnormalities	Refs.
Fmr1 KO	• Decreased deep nuclei volume • Reduced number of neurons in deep nuclei • Increased astrocytes in deep nuclei • Reduced volume of cerebellar cortex • Reduced cerebellar volume during development • Elongated spines in PC	• Deficits in eyeblink conditioning • Altered parallel fibers-PC synapses	[57–60]
Fmr1-PC cKO	• Reduced cerebellar volume • Cellular loss deep nuclei • Elongated spines in PC	• Deficits in eyeblink conditioning • Altered parallel fibers-PC synapses	[60]
TSC1-PC cKO +/−	• Abnormal spine density • Neurodegeneration of PC	• Decreased PC excitability • Deficits in eyeblink conditioning	[61, 62]
TSC1-PC cKO −/−	• Neurodegeneration of PC starting at 2 months	• Decreased PC excitability • Deficits in eyeblink conditioning	[61, 62]
TSC2-PC cKO+/−	• Increased size of PC and apoptosis	N/A	[63]
SHANK2 KO	N/A	• Decreased PC plasticity • Altered parallel fibers-PC synapses	[64]
SHANK3 ΔC	• Decreased density of PC • Lower spine density in PC • Reduced complexity of dendritic tree in PC	• Deficits in eyeblink conditioning	[62]

Table 1.
Cerebellar abnormalities in mouse models of ASD.

lobules are also abnormal in non-syndromic ASD [57]. Moreover, positive correlations between the size of the posterior vermis and several subscales of the autism behavior checklist, a list of nonadaptive behaviors that represent an individual's challenges to respond appropriately to daily life situations, in persons with FXS, have been reported [58]. On a different note, recent postmortem work has shown reductions in FMRP in cerebella and frontal cortices of subjects with autism who do not carry a mutation for FXS [59].

The Fmr1 knockout (KO) mouse is a validated and widely studied animal model of ASD. By means of high-resolution MRI imaging to study brain structure, the cerebellum in Fmr1 KO mice was found to show significant volume alterations comparing with wild-type controls. Specifically, the deep cerebellar nuclei, which transfer the output of the cerebellar cortex to the thalamus and cerebral cortex, were smaller in Fmr1 mice. Moreover, this reduced volume was accompanied by loss of neurons and increase in astrocytes, as measured by immunohistochemical techniques [60]. Later, the same authors also identified a volume reduction in the cerebellar cortex in these mice [67]. Further, a detailed study by diffusion tensor imaging of postnatal development in the Fmr1 KO mouse showed reduced cerebellar volume in the first few weeks after birth [68]. In addition to the full KO, a conditional Fmr1 KO mouse (Fmr1 cKO) has also been generated. Deletion of Fmr1 specifically in PC leads to several structural and functional abnormalities in the cerebellum also seen in the full Fmr1 KO, such as a reduction of cerebellar volume, cellular loss in the cerebellar nuclei, and longer spines in PC. Functionally, an enhanced LTD induction at the parallel fiber synapses that innervate these spines is seen [63]. In addition, both Fmr1 KO and PC-Fmr1 cKO mice present alterations in classical delay eyeblink conditioning, a Pavlovian associative learning where subjects learn to execute an appropriately timed eyeblink in response to a previously neutral conditioning stimulus in which the cerebellum plays a key role. Specifically, the percentage of conditioned responses and their peak amplitude and peak velocity were reduced. [63]. Interestingly, FXS patients were found to display the same cerebellar deficits in eyeblink conditioning as mutant mice [63]. Together these results suggest that the deficits in eyeblink conditioning are likely due to the loss of FMRP in PC.

5.2 Tuberous sclerosis (TSC1/TSC2 genes)

Tuberous sclerosis (TS) is another rare syndromic disorder associated with ASD. TS is characterized by the development of non-cancerous tumors in the brain and other organs leading to neurological symptoms such as developmental delay, epilepsy, and ASD. It is an autosomal dominant condition caused by mutations in either the TSC1 or TSC2 genes, which form a tumor suppressor complex involved in the regulation of the mTOR signaling pathway. Approximately 40% of TS patients are co-diagnosed with ASD, and, interestingly, those with cerebellar lesions have been found to have a more severe ASD diagnosis [61]. Further, PC loss has been found in postmortem cerebellum samples from TSC patients [62].

Several lines of mutant TSC mice have been generated and studied in detail. A mutant mouse in which the Tsc2 gene was selectively deleted from PCs starting at postnatal day 6 was generated to mimic patients with one nonfunctioning TSC2 allele [62]. The haploinsufficiency of TSC2 caused a progressive increase in PC cell size and subsequent death from apoptosis. TSC2-null PCs showed increased endoplasmic reticulum and oxidative stress, which were rescued by treatment with the mTOR inhibitor rapamycin. In a subsequent study, the authors reported that PC-TSC2-haploinsufficient mice showed social deficits and repetitive behaviors [64]. These observations indicate that selective loss of TSC2 in PCs in a TSC2-haploinsufficient background is enough to lead to autistic-like behavioral deficits.

A conditional PC-TSC1 KO mouse has also been generated. Both heterozygous and homozygous losses of TSC1 in mouse cerebellar PCs result in autistic-like behaviors, including abnormal social interaction and repetitive behavior and vocalizations, in addition to decreased PC excitability. Similar to TSC2 mutants, treatment of TSC1 mutant mice with the mTOR inhibitor, rapamycin, prevented the pathological and behavioral deficits. Strikingly, PC-TSC1 homozygous mice, but not PC-TSC1 heterozygous mice, showed PC loss at 2 months of age. The fact that PC-TSC1 heterozygous mice showed autistic symptoms and PC excitability alterations in the absence of PC loss suggests that the decrease in PC excitability is likely driving the phenotype [69]. Further, this model has also been reported to show deficits in eyeblink conditioning; they specifically show lower percentage of conditioned response in this test [70]. These findings demonstrate new roles for TSC1/TSC2 in PC function and define a molecular basis for a cerebellar contribution to cognitive disorders such as autism.

5.3 Phelan-McDermid syndrome (SHANK3 gene)

Phelan-McDermid syndrome (PMS) is due to heterozygous chromosome 22q13 deletions and is often co-diagnosed with ASD. The clinical manifestations of PMS include global developmental delay/intellectual disability and absent or delayed speech [71]. Although the deletion encompasses numerous genes, a good candidate that could account for ASD symptoms is SHANK3, a gene within which mutations have independently been associated with non-syndromic ASD. ASD patients with SHANK3 deletions are also known to have severe core symptoms and mental disabilities [72]. Although to our knowledge there is no data addressing the effect of SHANK3 mutations on cerebellar anatomy and function, recent research suggests that mutations in *SHANK3* may be related to cerebellar abnormalities. SHANK3 is highly expressed in cerebellar granule cells [73] and has been suggested to play a role in the recruitment of axon terminals to cerebellar granule cell dendrites [74]. In addition, cerebellar vermis hypoplasia has been found in patients with PMS, suggesting that SHANK3 may be involved in cerebellar development [75].

Multiple mouse lines with SHANK3 mutations exist, and several display behaviors analogous to the core symptoms of autism, including isoform-specific SHANK3B KO [76], SHANK3 (Δexons4–9) deleting major isoforms of the gene [77], and SHANK3 (ΔC), deleting the C-terminal region of the gene [78]. SHANK3 (ΔC) mice present a decreased density of PC compared to controls [70], and they show deficits in the eyeblink conditioning task, showing lower percentage of conditioning response and a delay in the response latency [70].

Interestingly, mutations in other proteins from the SHANK family, such as SHANK2, have been also linked to ASD. A KO mouse for SHANK2 shows alterations in social and repetitive behaviors and presents changes in PC electrophysiological characteristics, such as decreased intrinsic PC plasticity, synaptic strength at the PC-parallel fiber synapse, and enhanced inhibitory input into PC. Further PC-specific SHANK2 KO replicated these findings, arguing for a cerebellar role in autistic-like behaviors [79].

6. Conclusion

The cerebellum has been recently indicated as a key structure not only for sensorimotor control but also for language, social cognition, and emotion, via its extensive connections with cortical areas. In the present work, we aimed to provide an up-to-date overview of current findings on cerebellar involvement in the pathophysiology of ASD. Anatomical studies report cerebellar abnormalities in postmortem brain tissue from autistic individuals, neuroimaging studies indicate abnormal cerebellar activation when performing social paradigms, and animal models of monogenic forms of autism converge on the cerebellum as one of the common sites of abnormalities. The cerebellum represents an emerging field of interest for ASD research, based on the hypothesis that ASD is a connectivity disorder and cerebellar dysfunction could impact other brain areas within the social network, leading to the core ASD symptoms. Although the literature in this new field is at a very early stage, based on the presented data, future studies should not exclude the cerebellum in analyses of structural and functional differences in ASD.

Acknowledgements

This work was supported by the following grants: Spanish Ministry of Economy and Competitiveness/European Regional Development Fund (MINECO/FEDER grant SAF2015-64163-R). OP is a Ramon y Cajal Fellow (RYC-2013-12558), MF is supported by a predoctoral fellowship from the Spanish Ministry of Economy and Competitiveness MINECO (BES-2016-078420), and TS-A is supported by a predoctoral fellowship from the Basque Government.

Author details

Marta Fernández, Teresa Sierra-Arregui and Olga Peñagarikano*
Department of Pharmacology, University of the Basque Country (UPV/EHU), Spain

*Address all correspondence to: olga.penagarikano@ehu.eus

References

[1] Association AP. Diagnostic and Statistical Manual of Mental Disorders. 5th ed. Washington DC: American Psychiatric Publishing; 2013

[2] Adolphs R. The neurobiology of social cognition. Current Opinion in Neurobiology. 2001;**11**(2):231-239

[3] Baio J, Wiggins L, Christensen DL, Maenner MJ, Daniels J, Warren Z, et al. Prevalence of autism Spectrum disorder among children aged 8 years—Autism and developmental disabilities monitoring network, 11 sites, United States, 2014. MMWR Surveillance Summaries. 2018;**67**(6):1-23

[4] Geschwind DH. Advances in autism. Annual Review of Medicine. 2009;**60**:367-380

[5] Sandin S, Lichtenstein P, Kuja-Halkola R, Hultman C, Larsson H, Reichenberg A. The heritability of autism spectrum disorder. Journal of the American Medical Association. 2017;**318**(12):1182-1184

[6] Berg JM, Geschwind DH. Autism genetics: Searching for specificity and convergence. Genome Biology. 2012;**13**(7):247

[7] Bolte S, Girdler S, Marschik PB. The contribution of environmental exposure to the etiology of autism spectrum disorder. Cellular and Molecular Life Sciences. 2018;**76**(7):1275-1297

[8] Baron-Cohen S, Ring HA, Wheelwright S, Bullmore ET, Brammer MJ, Simmons A, et al. Social intelligence in the normal and autistic brain: An fMRI study. The European Journal of Neuroscience. 1999;**11**(6):1891-1898

[9] Brothers L, Ring B, Kling A. Response of neurons in the macaque amygdala to complex social stimuli. Behavioural Brain Research. 1990;**41**(3):199-213

[10] Baez-Mendoza R, Schultz W. The role of the striatum in social behavior. Frontiers in Neuroscience. 2013;7:233

[11] Kennedy DP, Adolphs R. The social brain in psychiatric and neurological disorders. Trends in Cognitive Sciences. 2012;**16**(11):559-572

[12] Courchesne E, Mouton PR, Calhoun ME, Semendeferi K, Ahrens-Barbeau C, Hallet MJ, et al. Neuron number and size in prefrontal cortex of children with autism. Journal of the American Medical Association. 2011;**306**(18):2001-2010

[13] Courchesne E, Pierce K. Why the frontal cortex in autism might be talking only to itself: Local over-connectivity but long-distance disconnection. Current Opinion in Neurobiology. 2005;**15**(2):225-230

[14] Schumann CM, Barnes CC, Lord C, Courchesne E. Amygdala enlargement in toddlers with autism related to severity of social and communication impairments. Biological Psychiatry. 2009;**66**(10):942-949

[15] Kleinhans NM, Richards T, Weaver K, Johnson LC, Greenson J, Dawson G, et al. Association between amygdala response to emotional faces and social anxiety in autism spectrum disorders. Neuropsychologia. 2010;**48**(12):3665-3670

[16] Chevallier C, Kohls G, Troiani V, Brodkin ES, Schultz RT. The social motivation theory of autism. Trends in Cognitive Sciences. 2012;**16**(4):231-239

[17] Dichter G, Adolphs R. Reward processing in autism: A thematic series. Journal of Neurodevelopmental Disorders. 2012;**4**(1):20

[18] Schmahmann JD, Caplan D. Cognition, emotion and the cerebellum. Brain. 2006;**129**(Pt 2): 290-292

[19] Wegiel J, Flory M, Kuchna I, Nowicki K, Ma SY, Imaki H, et al. Stereological study of the neuronal number and volume of 38 brain subdivisions of subjects diagnosed with autism reveals significant alterations restricted to the striatum, amygdala and cerebellum. Acta Neuropathologica Communications. 2014;**2**:141

[20] Mosconi MW, Wang Z, Schmitt LM, Tsai P, Sweeney JA. The role of cerebellar circuitry alterations in the pathophysiology of autism spectrum disorders. Frontiers in Neuroscience. 2015;**9**:296

[21] Geschwind DH, Levitt P. Autism spectrum disorders: Developmental disconnection syndromes. Current Opinion in Neurobiology. 2007;**17**(1):103-111

[22] Crawley JN. Behavioral phenotyping strategies for mutant mice. Neuron. 2008;**57**(6):809-818

[23] Rubenstein JL, Merzenich MM. Model of autism: Increased ratio of excitation/inhibition in key neural systems. Genes, Brain, and Behavior. 2003;**2**(5):255-267

[24] Yizhar O, Fenno LE, Prigge M, Schneider F, Davidson TJ, O'Shea DJ, et al. Neocortical excitation/inhibition balance in information processing and social dysfunction. Nature. 2011;**477**(7363):171-178

[25] Schultz W. Updating dopamine reward signals. Current Opinion in Neurobiology. 2013;**23**(2):229-238

[26] Gunaydin LA, Grosenick L, Finkelstein JC, Kauvar IV, Fenno LE, Adhikari A, et al. Natural neural projection dynamics underlying social behavior. Cell. 2014;**157**(7):1535-1551

[27] Carta I, Chen CH, Schott AL, Dorizan S, Khodakhah K. Cerebellar modulation of the reward circuitry and social behavior. Science. 2019;**363**(6424): pii: eaav0581

[28] Fernandez M, Mollinedo-Gajate I, Penagarikano O. Neural circuits for social cognition: Implications for autism. Neuroscience. 2018;**370**:148-162

[29] Raymond JL, Lisberger SG, Mauk MD. The cerebellum: A neuronal learning machine? Science. 1996;**272**(5265):1126-1131

[30] Stoodley CJ, Schmahmann JD. Evidence for topographic organization in the cerebellum of motor control versus cognitive and affective processing. Cortex. 2010;**46**(7):831-844

[31] Bostan AC, Strick PL. The basal ganglia and the cerebellum: Nodes in an integrated network. Nature Reviews. Neuroscience. 2018;**19**(6):338-350

[32] Strick PL, Dum RP, Fiez JA. Cerebellum and nonmotor function. Annual Review of Neuroscience. 2009;**32**:413-434

[33] Stoodley CJ, Schmahmann JD. The cerebellum and language: Evidence from patients with cerebellar degeneration. Brain and Language. 2009;**110**(3):149-153

[34] Jack A, Englander ZA, Morris JP. Subcortical contributions to effective connectivity in brain networks supporting imitation. Neuropsychologia. 2011;**49**(13):3689-3698

[35] Jack A, Pelphrey KA. Neural correlates of animacy attribution include neocerebellum in healthy adults. Cerebral Cortex. 2015;**25**(11):4240-4247

[36] Van Overwalle F, Baetens K, Marien P, Vandekerckhove M. Social cognition and the cerebellum: A meta-analysis of over 350 fMRI studies. NeuroImage. 2014;**86**:554-572

[37] Van Overwalle F, Baetens K, Marien P, Vandekerckhove M. Cerebellar areas dedicated to social cognition? A comparison of meta-analytic and connectivity results. Social Neuroscience. 2015;**10**(4):337-344

[38] Rogers TD, McKimm E, Dickson PE, Goldowitz D, Blaha CD, Mittleman G. Is autism a disease of the cerebellum? An integration of clinical and pre-clinical research. Frontiers in Systems Neuroscience. 2013;**7**:15

[39] D'Mello AM, Crocetti D, Mostofsky SH, Stoodley CJ. Cerebellar gray matter and lobular volumes correlate with core autism symptoms. NeuroImage: Clinical. 2015;**7**:631-639

[40] Wang SS, Kloth AD, Badura A. The cerebellum, sensitive periods, and autism. Neuron. 2014;**83**(3):518-532

[41] Fatemi SH, Aldinger KA, Ashwood P, Bauman ML, Blaha CD, Blatt GJ, et al. Consensus paper: Pathological role of the cerebellum in autism. Cerebellum. 2012;**11**(3):777-807

[42] Skefos J, Cummings C, Enzer K, Holiday J, Weed K, Levy E, et al. Regional alterations in purkinje cell density in patients with autism. PLoS One. 2014;**9**(2):e81255

[43] Courchesne E, Karns CM, Davis HR, Ziccardi R, Carper RA, Tigue ZD, et al. Unusual brain growth patterns in early life in patients with autistic disorder: An MRI study. Neurology. 2001;**57**(2):245-254

[44] Bolduc ME, du Plessis AJ, Sullivan N, Guizard N, Zhang X, Robertson RL, et al. Regional cerebellar volumes predict functional outcome in children with cerebellar malformations. Cerebellum. 2012;**11**(2):531-542

[45] Riva D, Annunziata S, Contarino V, Erbetta A, Aquino D, Bulgheroni S. Gray matter reduction in the vermis and CRUS-II is associated with social and interaction deficits in low-functioning children with autistic spectrum disorders: A VBM-DARTEL study. Cerebellum. 2013;**12**(5):676-685

[46] Bailey A, Luthert P, Dean A, Harding B, Janota I, Montgomery M, et al. A clinicopathological study of autism. Brain. 1998;**121**(Pt 5):889-905

[47] Vargas DL, Nascimbene C, Krishnan C, Zimmerman AW, Pardo CA. Neuroglial activation and neuroinflammation in the brain of patients with autism. Annals of Neurology. 2005;**57**(1):67-81

[48] Yip J, Soghomonian JJ, Blatt GJ. Decreased GAD67 mRNA levels in cerebellar Purkinje cells in autism: Pathophysiological implications. Acta Neuropathologica. 2007;**113**(5):559-568

[49] Yip J, Soghomonian JJ, Blatt GJ. Increased GAD67 mRNA expression in cerebellar interneurons in autism: Implications for Purkinje cell dysfunction. Journal of Neuroscience Research. 2008;**86**(3):525-530

[50] Hegarty JP 2nd, Weber DJ, Cirstea CM, Beversdorf DQ. Cerebro-cerebellar functional connectivity is associated with cerebellar excitation-inhibition balance in autism spectrum disorder. Journal of Autism and Developmental Disorders. 2018;**48**(10):3460-3473

[51] Khan AJ, Nair A, Keown CL, Datko MC, Lincoln AJ, Muller RA. Cerebro-cerebellar resting-state functional connectivity in children and adolescents with autism spectrum disorder. Biological Psychiatry. 2015;**78**(9):625-634

[52] Verly M, Verhoeven J, Zink I, Mantini D, Peeters R, Deprez S, et al. Altered functional connectivity of the language network in ASD: Role of classical language areas and cerebellum. NeuroImage: Clinical. 2014;**4**:374-382

[53] Padmanabhan A, Lynn A, Foran W, Luna B, O'Hearn K. Age related changes in striatal resting state functional connectivity in autism. Frontiers in Human Neuroscience. 2013;**7**:814

[54] Mostofsky SH, Powell SK, Simmonds DJ, Goldberg MC, Caffo B, Pekar JJ. Decreased connectivity and cerebellar activity in autism during motor task performance. Brain. 2009;**132**(Pt 9):2413-2425

[55] Jack A, Morris JP. Neocerebellar contributions to social perception in adolescents with autism spectrum disorder. Developmental Cognitive Neuroscience. 2014;**10**:77-92

[56] Kana RK, Maximo JO, Williams DL, Keller TA, Schipul SE, Cherkassky VL, et al. Aberrant functioning of the theory-of-mind network in children and adolescents with autism. Molecular Autism. 2015;**6**:59

[57] Kaufmann WE, Cooper KL, Mostofsky SH, Capone GT, Kates WR, Newschaffer CJ, et al. Specificity of cerebellar vermian abnormalities in autism: A quantitative magnetic resonance imaging study. Journal of Child Neurology. 2003;**18**(7):463-470

[58] Gothelf D, Furfaro JA, Hoeft F, Eckert MA, Hall SS, O'Hara R, et al. Neuroanatomy of fragile X syndrome is associated with aberrant behavior and the fragile X mental retardation protein (FMRP). Annals of Neurology. 2008;**63**(1):40-51

[59] Fatemi SH, Folsom TD. Dysregulation of fragile X mental retardation protein and metabotropic glutamate receptor 5 in superior frontal cortex of individuals with autism: A postmortem brain study. Molecular Autism. 2011;**2**:6

[60] Ellegood J, Pacey LK, Hampson DR, Lerch JP, Henkelman RM. Anatomical phenotyping in a mouse model of fragile X syndrome with magnetic resonance imaging. NeuroImage. 2010;**53**(3):1023-1029

[61] Eluvathingal TJ, Behen ME, Chugani HT, Janisse J, Bernardi B, Chakraborty P, et al. Cerebellar lesions in tuberous sclerosis complex: Neurobehavioral and neuroimaging correlates. Journal of Child Neurology. 2006;**21**(10):846-851

[62] Reith RM, Way S, McKenna J 3rd, Haines K, Gambello MJ. Loss of the tuberous sclerosis complex protein tuberin causes Purkinje cell degeneration. Neurobiology of Disease. 2011;**43**(1):113-122

[63] Koekkoek SK, Yamaguchi K, Milojkovic BA, Dortland BR, Ruigrok TJ, Maex R, et al. Deletion of FMR1 in Purkinje cells enhances parallel fiber LTD, enlarges spines, and attenuates cerebellar eyelid conditioning in fragile X syndrome. Neuron. 2005;**47**(3):339-352

[64] Reith RM, McKenna J, Wu H, Hashmi SS, Cho SH, Dash PK, et al. Loss of TSC2 in Purkinje cells is associated with autistic-like behavior in a mouse model of tuberous sclerosis complex. Neurobiology of Disease. 2013;**51**:93-103

[65] Penagarikano O, Mulle JG, Warren ST. The pathophysiology of fragile X syndrome. Annual Review of Genomics and Human Genetics. 2007;**8**:109-129

[66] Greco CM, Navarro CS, Hunsaker MR, Maezawa I, Shuler JF, Tassone F, et al. Neuropathologic features in the hippocampus and cerebellum of three older men with fragile X syndrome. Molecular Autism. 2011;**2**(1):2

[67] Ellegood J, Anagnostou E, Babineau BA, Crawley JN, Lin L, Genestine M, et al. Clustering autism: Using neuroanatomical differences in 26 mouse models to gain insight into the heterogeneity. Molecular Psychiatry. 2015;**20**(1):118-125

[68] Pacey LK, Xuan IC, Guan S, Sussman D, Henkelman RM, Chen Y, et al. Delayed myelination in a mouse model of fragile X syndrome. Human Molecular Genetics. 2013;**22**(19):3920-3930

[69] Tsai PT, Hull C, Chu Y, Greene-Colozzi E, Sadowski AR, Leech JM, et al. Autistic-like behaviour and cerebellar dysfunction in Purkinje cell TSC1 mutant mice. Nature. 2012;**488**(7413):647-651

[70] Kloth AD, Badura A, Li A, Cherskov A, Connolly SG, Giovannucci A, et al. Cerebellar associative sensory learning defects in five mouse autism models. eLife. 2015;**4**:e06085

[71] Phelan K, McDermid HE. The 22q13.3 deletion syndrome (Phelan-McDermid syndrome). Molecular Syndromology. 2012;**2**(3-5):186-201

[72] Chen BY, Zou XB, Zhang J, Deng HZ, Li JY, Li LY, et al. Copy-number variations of SHANK3 and related clinical phenotypes in children with autism. Zhonghua Er Ke Za Zhi. 2011;**49**(8):607-611

[73] Beri S, Tonna N, Menozzi G, Bonaglia MC, Sala C, Giorda R. DNA methylation regulates tissue-specific expression of SHANK3. Journal of Neurochemistry. 2007;**101**(5):1380-1391

[74] Roussignol G, Ango F, Romorini S, Tu JC, Sala C, Worley PF, et al. SHANK expression is sufficient to induce functional dendritic spine synapses in aspiny neurons. The Journal of Neuroscience. 2005;**25**(14):3560-3570

[75] Aldinger KA, Kogan J, Kimonis V, Fernandez B, Horn D, Klopocki E, et al. Cerebellar and posterior fossa malformations in patients with autism-associated chromosome 22q13 terminal deletion. American Journal of Medical Genetics. Part A. 2013;**161A**(1):131-136

[76] Peca J, Feliciano C, Ting JT, Wang W, Wells MF, Venkatraman TN, et al. SHANK3 mutant mice display autistic-like behaviours and striatal dysfunction. Nature. 2011;**472**(7344):437-442

[77] Wang X, McCoy PA, Rodriguiz RM, Pan Y, Je HS, Roberts AC, et al. Synaptic dysfunction and abnormal behaviors in mice lacking major isoforms of SHANK3. Human Molecular Genetics. 2011;**20**(15):3093-3108

[78] Kouser M, Speed HE, Dewey CM, Reimers JM, Widman AJ, Gupta N, et al. Loss of predominant SHANK3 isoforms results in hippocampus-dependent impairments in behavior and synaptic transmission. The Journal of Neuroscience. 2013;**33**(47):18448-18468

[79] Peter S, Ten Brinke MM, Stedehouder J, Reinelt CM, Wu B, Zhou H, et al. Dysfunctional cerebellar Purkinje cells contribute to autism-like behaviour in SHANK2-deficient mice. Nature Communications. 2016;**7**:12627

Section 3

Behavioral Neuroscience in Adulthood - Examples of Hot Topics in Psychiatry and Addiction

Chapter 7

Treatment-Induced Brain Plasticity in Psychiatric Disorders

Maria Uscinska, Andrea Polla Mattiot and Silvio Bellino

Abstract

In tandem with a better-informed neurobiological model of mental illness, psychiatry has progressively been shaped into its current state of clinical neuroscience. The traditional dichotomy of organic versus endogenous mental disorders has been replaced by the growing recognition that all changes in mental processes are accompanied by changes in structures or functions of the brain. Thus, all psychiatric interventions are deemed to have a biopsychosocial nature, whereby drugs in addition to their effect on the brain have a psychological effect, and psychotherapies beyond their psychological effects may alter the brain. In this view, the ultimate goal of any psychiatric treatment is to induce neural plasticity in a manner that restores the full original function and potential of the injured brain. Herein present chapter gives an insight into how evidence-based treatments achieve their therapeutic effects on the level of cerebral reorganization across a host of psychiatric disorders. The main theme of this work is the posited mechanism of neuroplasticity on neural-systems level for each treatment modality.

Keywords: therapy, drugs, treatment, psychosis, schizophrenia, depression, disorder, antidepressant, antipsychotic, neuroimaging, fMRI, cognitive, plasticity, mood stabilizers, neural correlates

1. Neural parameters of therapeutic change

Mechanisms of neuroplasticity constitute fundamental processes behind learning and memory, that determine the ability of neuronal systems to incorporate novel environmental stimuli and to make appropriate adaptive response. Delineating cerebral processes of recovery from an insult to the brain holds promise for developing more refined and novel treatment modalities to target specific areas of pathology. Functional neuroimaging studies provide a mean to characterize changes in brain function related to psychiatric interventions. Well-established in indexing biomarkers of psychiatric disorders, novel neuroimaging techniques are now used to depict patterns of neural plasticity mediating post-treatment amelioration of symptoms. Various modalities provide indices of brain activity by measuring cerebral blood flow or glucose metabolism including functional magnetic resonance imaging (fMRI), 18fluorodeoxyglucose positron emission tomography (FDG-PET), and 99mtechnetium hexamethylpropyleneamineoxime single photon emission computed tomography (99mTc-HMPAO SPECT) [see ref. 1 for a detailed review] One powerful imaging modality that has significantly advanced our knowledge in this field is the task-based functional magnetic resonance imaging (t-fMRI).

It consists of a paradigm defined by a functional measurement including a stimulation adjusted to the brain area under investigation. The subject is required to perform a defined motor or sensorimotor, language or another cognitive or visual tasks in the MRI scanner while typically GRE T2*-weighed echo planner images (EDI) are rapidly acquired [for a more in-depth description of fMRI see ref. 1]. Local changes in cerebral blood flow (CBF) during task execution relative to resting state are used to infer brain regions/networks functionally involved in specific tasks. To ultimately determine the specificity and amount of therapy-induced neuroplasticity, multiple pre-, and post- therapy scans are compared against activity pattern changes in other active treatment groups and a no-treatment waiting-list group [2]. With this in mind, the next section follows with an overview of main findings associated with intervention-induced neuroplasticity and their interpretations.

2. Putative neuroplastic mechanisms of pharmacotherapy

Pharmacotherapy constitutes first-line treatment modality for majority of psychiatric disorders, and various theories exist as to how drug-induced neurochemical changes reverse different psychiatric symptoms. The posited purely neurotransmitter-based mechanism of action postulates either increased or reduced synaptic concentration of a target neurotransmitter that is implicated in a given disorder. This model is challenged by disjunction in the timescale of the onset of neurochemical versus therapeutic effects, wherein the potentiation or attenuation of neurotransmitter function often occurs within hours of administration and the clinical improvement is typically seen days or weeks after [3]. In quest of new rapid-acting agents, contemporary approaches to understanding of drug action focus on the role of adaptive neuroplastic processes that correlate in time with the onset of clinical improvement, hence are hypothesized to represent a more direct treatment target.

2.1 Antidepressant drugs and mechanisms of neuroplasticity

Current national and international guidelines recommend serotonin reuptake inhibitors (SSRIs) as first-line treatment for most patients with major depression, and the use of serotonin—norepinephrine reuptake inhibitors (SNRI) in patients resistant to the former [4, 5]. Although novel, better tolerated and more selective inhibitors of serotonin and norepinephrine reuptake are continuously being developed, the efficacy of tricyclic antidepressants such as amitriptyline for severe depression, has never been surpassed [6].

Most of currently licensed antidepressants act to enhance monoamine neurotransmission, where they are believed to achieve therapeutic effects by increasing availability of serotonin or/and norepinephrine, at least initially [7]. On the neural level, antidepressants normalize aberrant neural activity patterns underlying negative bias in affective information processing, posited to play central role in the etiology and maintenance of depressed state [8]. Thus, antidepressants were shown to attenuate hyperactivity in limbic areas of the brain (amygdala, insula, anterior cingulate), and enhance regulatory activity in the dorsolateral and medial prefrontal cortex as measured by functional magnetic resonance imaging [9, 10]. It was demonstrated that 7 days treatment with SSRI, citalopram, SNRI, and reboxetine reversed abnormal patterns of neural response to affective information, and induced a similar direction of change in healthy individuals [11, 12]. Noteworthy, short-term SSRI administration normalized amygdala hyperactivity in response to negative emotional stimuli prior to clinical changes in mood ratings in placebo-controlled studies [13]. These findings allow to speculate that treatment-induced early reversal of negative emotional bias sets the scene for therapeutic recovery over time by reducing the influence of this key maintaining factor [14].

2.2 Antipsychotic drugs and mechanisms of neuroplasticity

Antipsychotic medication is the mainstay of effective management of psychosis where schizophrenia is the most prevalent among psychotic disorders. Most of what we know about antipsychotic drugs action is at the receptor level, where abnormalities in neurotransmission constitute either an excess or a deficiency of neurotransmitters, including dopamine, serotonin, and glutamate. Therein first-generation drugs act as antagonists of dopamine D2 receptors and target most positive symptoms such as hallucinations and delusions. The receptor-binding profile of second-generation drugs extends beyond D2 affinity antagonism to other neuroreceptors including serotonin 5-HT2A in the frontal lobe, thus accounting for superior efficacy of these drugs in the pathophysiology of negative symptoms and cognitive disorganization [15, 16]. Overall, treatment response has been shown to be associated to the level of D2 occupancy, which is the target of all currently licensed antipsychotics [17]. To delineate therapeutic mechanism of clinically effective drugs beyond receptor level, research has focused on neural systems effects before and after pharmacotherapy in medication-naïve patients with first-episode psychosis. Functional MRI studies revealed pre-treatment functional alterations within frontostriatal circuitry, marked by patterns of hypoactivity within the dorsolateral/medial prefrontal cortex coupled with hyperactivity in the hippocampus and striatum [18–20]. Thus, aberrant frontostriatal circuitry might represent a potential system- level mechanism of psychosis and a candidate for treatment target with antipsychotics. Post-treatment findings lend some evidence to validate this model, showing increases in task-related frontal cortical activation in patients who underwent 12 weeks of quetiapine fumarate treatment compared to a drug-naive group [21, 22], and in a small group of patients with schizophrenia medicated with risperidone [23]. A similar study on cortical structure and function alterations within 1 year of psychosis onset in unmedicated schizophrenia patients versus patients under short-term therapy with atypical antipsychotics revealed a more complex relationship [20]. Although the treatment was associated with enhanced cognitive control and increased prefrontal, middle temporal, parietal, and occipital activity, it also revealed post-treatment prefrontal cortical thinning in the treatment group. The mechanism by which antipsychotics are associated with the loss of gray matter remains unclear, however neuroinflammatory models posit elevations in proinflammatory cytokine levels [24], microglia activation [25], and increased extracellular volume in white and gray matter [26]. Thus, the study adds to the growing literature on therapeutic mechanisms of antipsychotics, mediated by normalization of aberrant frontal cortical function, and suggests that caution must be exercised in interpreting neuroanatomical changes as being potentially deleterious to brain function.

2.3 Mood stabilizers and mechanisms of neuroplasticity

Lithium and anticonvulsants with mood-stabilizing properties (lamotrigine, valproate) constitute first-line drug treatment for episodes of depression and mania with variable inter-episode remission [27–29]. Whilst different compounds may differentially target specific facets of bipolar disorders, lithium is effective for all phases including acute depression [30]. On the neural level, functional imaging studies consistently point to pre-treatment frontolimbic dysfunction during cognitive control and emotion-paradigms in bipolar disorder patients [31–33]. Thus, abnormal emotion regulation and impaired cognition might be attributed to interference in cognitive control within medial prefrontal cortex though overactivity in subcortical structures (amygdala, ACC, insula), involved in emotion generation and appraisal. Findings of mood stabilizers-induced neural plasticity yield less consistent results due to methodological limitations that make it difficult to draw firm conclusions. Whilst some studies find no significant effects of pharmacotherapy upon functional measures of cerebral reorganization in bipolar patients [34–41] others reported increased task-related prefrontal cortical activity coupled with normalized subcortical limbic activity during emotional processing [38, 39, 42, 43]. Typically, individuals recruited in these studies are able to tolerate medication withdrawal and washout, and therefore are likely to have a milder form of the disorder. Given that it is not clinically feasible to withdraw all patients with bipolar disorder from medication, individuals with a more severe form of the disorder are likely to be underrepresented in many studies and therefore findings might not be generalizable to the most at-need of new treatments group.

3. Putative neuroplastic mechanisms of psychotherapy

Although studies of neural parameters of therapeutic change under psychotherapy are under-represented relative to analogous studies of medications, emerging literature support the thesis that changes in affect, cognition and behavior mediated by psychotherapy have demonstrable neuroplastic underpinnings. Since the call for more neuroscientifically informed approaches to psychotherapy [44], studies have elucidated neural mechanism of psychotherapy-induced changes in brain activity profiles across a range of psychiatric disorders.

3.1 Cognitive behavioral therapy and mood disorders

Psychotherapy processes appear to target maladaptive cognitive and emotional patterns by engaging their biological analogues that are responsive to a discrete mode of treatment [45]. One salient example involves re-appraisal technique under cognitive behavioral therapy (CBT) for depression, where patients are invited to re-interpret their negative perceptions of unpleasant occurrences in a more positive light. Mood ratings before and after re-thinking negative events revealed improved positive affect, mediated by elevated activity in dorsolateral and dorsomedial PFC coupled with decreased activity in the amygdala and orbitofrontal cortex [46]. To delineate CBT- induced mechanism of neuroplasticity in depression, FDG-PET scans before and after psychotherapy relative to paroxetine treatment were acquired from patients instructed to 'avoid ruminating on any one topic' during scanning [47]. Although efficacy of both treatments was comparable, differential activity patterns emerged in frontal and limbic regions, implying that medication and psychotherapy might achieve their therapeutic effects in different ways. Whilst CBT was associated with decreased metabolism in multiple frontal regions including the dorsolateral PFC together with increased activity in the hippocampus, parahippocampal gyrus, and dorsal cingulate gyrus, paroxetine- induced increased PFC metabolism, and decrease in hippocampal, parahippocampal, posterior cingulate and ventral subgenual cingulate activity. This modality-specific mechanism of neuroplasticity posits that CBT exerts 'top- down' changes in cognitive processing in favor of engaging ventral and limbic regions, which mediate attention to personally salient stimuli, whereas antidepressant drugs prompt 'bottom-up' disengagement of ventral, frontal and limbic regions. Although this model runs counter the aforementioned emotion regulation model, divergent findings might result from using healthy subjects in the former study and patients with depression in the latter, invoking the notion that brain activation results from the interaction between underlying brain state and treatment modality [48].

In efforts to elucidate CBT- induced neural mechanism of anxiety disorders functional neuroimaging study examined pre-and-post CBT brain activity patterns in non-medicated patients with spider phobia and healthy subjects [49]. The former exhibited elevated activation in the parahippocampal gyrus and right dorsolateral PFC prior to the treatment, which was normalized with successful group CBT sessions focused on exposure therapy. Given that parahippocampal gyrus mediates contextual memory, authors suggested that after CBT less demand was placed on the dorsolateral PFC to construct a cognitive defense to the perceived threat. Moreover, a therapy- induced shift of activity to the ventral PFC was indexed, which might play a role in down-regulation of limbic activity and thereby dampening fear reaction. Collectively, these studies depict a neuroplastic model of cognitive behavioral therapy which posits altered engagement of dorsal prefrontal circuitry to down-regulate limbic and ventral prefrontal structures thereby improving affect in response to emotionally significant contexts.

3.2 Dialectic behavioral therapy and borderline personality disorder

Given that psychotherapy is the gold standard treatment modality for borderline personality disorder [50], extensive research efforts focused on measuring brain changes induced by specific modes of therapy. To date, dialectic behavioral therapy is the most researched, refined and evidenced-based therapy informed by a deficit model in self-regulation, distress tolerance and interpersonal skills, deemed to arise from transaction between highly sensitive individuals and invalidating environments [51, 52]. DBT purports to render individuals more mindful and able to manage relationships effectively by incorporating the concept of dialectics and strategy of validation into approach focused on skills acquisition and behavioral shaping.

Consistent with the skills deficit model of BPD, neuroimaging evidence supports that acquisition of affective control strategies under DBT balances neural substrates of emotion regulation. One salient example indexed neural activity alterations under re-appraisal and reported dampened insula and ACC activity, together with an enhanced connectivity of the latter to medial and superior frontal gyrus, superior temporal gyrus, and inferior parietal cortices [53]. Notably, treatment-induced increase in dorsal ACC activity during exposure to negative stimuli was associated with improvement self-reported BPD symptoms, suggesting a possible biomarker of improved affect regulation. In a similar study Winter et al. [54] set out to establish whether neural correlates of distraction might be amenable to a successful DBT. In this view, 31 BPD patients under constant medication were scanned before and after a 12- week residential DBT-based treatment while performing a distraction task. When compared to 15 BPD control patients under non-DBT-based treatment or no treatment at all, and 22 healthy participants, 16 DBT responders exhibited attenuated activity in the right inferior parietal lobe/supramarginal gyrus. Notably, this pattern of brain activity was correlated with improvement in self-reported borderline symptom severity (ZAN-BPD). Furthermore, treatment was associated with a reduction in the right perigenual ACC activity and increased activity in these regions during distraction in the context of aversive stimuli. These findings might reflect a shift from emotional to more cognitive processing in the context of aversive stimuli, thereby suggesting an improvement in emotional susceptibility under DBT.

Taken together aforementioned studies support that DBT processes target maladaptive emotional patterns by altering their biological analogues that are responsive to discrete cognitive strategies. DBT normalizes frontolimbic imbalances as part of the disturbed circuitry, which appear to mediate amelioration of BPD symptomatology. Caution must be exercised however, while interpreting results as medications may attenuate emotional responses in BPD patients [55], and giving combinations of drug subtypes makes it impossible to isolate the effect of a single agent.

3.3 IPT and depression

IPT is a short-term treatment that typically consists of 12–16 one-hour weekly sessions focused on improving interpersonal relationships. Drawing directly on identifiable issues between patients and therapists, it purports to instil the ability to make the necessary adjustments in interpersonal situations that will help to reduce symptoms of depression. Several imaging studies have examined biomarkers of cerebral reorganization induced by IPT relative to pharmacotherapy. One of them compared the effects of the former and venlafaxine (37.5 mg daily) on regional CBF using 99mTc-HMPAO SPECT in 28 drug-naive or drug-free patients with MDD [56]. Whilst comparative clinical improvements were mediated by elevated activity in the right basal ganglia in both treatment groups, patients in the IPT group also exhibited an increase in the right posterior cingulate activity. However, drawing firm conclusion from these findings is hampered by methodological issues as four patients with a strong preference for venlafaxine could choose the treatment, while one preferred IPT. Moreover, subjects in the latter evidenced greater striatal perfusion, potentially reflecting design limitation. Brief duration of IPT and relatively low dose of venlafaxine give rise to the possibility that both treatments were suboptimal, thereby underscoring the engagement of limbic and paralimbic recruitment in psychotherapy-induced changes reported in parallel research [56].

A similar study on the effects of IPT and paroxetine relative to healthy controls [57] reported results analogous to CBT effects described by Goldapple et al. [47]. Whilst treatment response in both groups was associated with an increase in metabolism in limbic and paralimbic regions (the right insula and left inferior temporal lobe) relative to controls, unlike CBT the effects of IPT were mediated by a decrease in dorsal and ventral prefrontal cortical metabolism. A follow-up study was set out to correlate treatment-mediated changes in brain activity patterns with amelioration in mood symptoms measured by the Hamilton Depression Rating Scale and the tension/anxiety and fatigue clusters of the Profile of Mood States [58]. A cohort of 39 patients under either paroxetine or IPT for MDD exhibited post-treatment reductions in ventral and dorsal frontal lobe metabolism, which was associated with improvements in the anxiety/somatization and psychomotor retardation symptom clusters. Unlike previous findings of negative correlation between activity in the dorsolateral PFC and improvement on global depression scores under CBT, in the present study alterations in dorsolateral PFC activity positively correlated with improvement in cognitive disturbance. These suggest that each treatment modality engages dorsolateral PFC function differently to achieve a specific therapeutic effect. While CBT appears to engage this region to attenuate 'over-thinking' in depression, IPT might induce it to improve general cognitive abilities.

3.4 Psychoeducation and euthymic bipolar disorder

Given that pharmacotherapy is often ineffective for treatment of residual depressive, dysthymic and dysphoric symptoms [59], researchers have shown interest in psychoeducation in targeting emotional and cognitive processes [60–63]. Psychoeducation is a treatment option for bipolar disorder focused on improving coping strategies to manage symptoms in everyday life, compliance with medication to prevent thymic relapses, quality of life and social functioning [64, 65]. Whilst wealth of research exists to support its efficacy in clinical symptoms improvement, less is known how therapeutic change is achieved on the level of neural functioning.

Favre and collaborates [66] set out to index neural processes before and after psychoeducation therapy in 16 euthymic bipolar patients (EBP) matched against 16 healthy subjects. Pre-treatment fMRI scans revealed reduced activity of cognitive control regions (bilateral inferior and left superior frontal gyri, right insula, right fusiform gyrus and bilateral occipital gyri) and elevated activity of emotion-related processing regions (bilateral hippocampus, parahippocampal gyri and the left middle temporal gyrus) in the treatment group. Thus, aberrant cognitive and emotion processing that characterize acute episodes in bipolar disorder appear to persist during euthymic phase. Post-treatment clinical improvement was mediated by increased activity of inferior frontal gyri and a pattern of decreased activity of right hippocampus and parahippocampal gyrus. These findings suggest that psychoeducation improves cognitive control by engaging prefrontal networks and normalizes generation of emotional responses by quieting activity within limbic networks.

3.5 Cognitive remediation therapy and schizophrenia

Cognitive remediation therapy (CRT) is an evidence-based treatment for neuropsychological deficits in memory, attention, executive function, social cognition or metacognition across a host of neuropsychiatric disorders [67–69]. There is a growing literature focused on neurobiological changes that mediate cognitive recovery under this type of intervention in patients with schizophrenia [70–73], mood disorders [74], mild cognitive impairment [75] and in healthy adults [76]. Majority of studies examined the effects of cognitive remediation on brain functioning in patients with schizophrenia and have amounted to several systematic reviews and meta-analyses [76–78]. Findings lend support to the frontal hypoactivation mechanism of cognitive impairment and suggest that cognitive remediation improves these networks efficiency. Most commonly reported areas of post-treatment amelioration in efficiency involved prefrontal and thalamic regions. Meusel and collaborates [73] set out to describe functional correlates of cognitive remediation in patients with bipolar disorder or depression versus healthy controls. Thirty eight subjects completed 10 weeks of treatment and were scanned (fMRI) during an n-back working memory task and a recollection memory task to investigate the potential for change within these networks. PRE-POST improvements correlated with functional activation in lateral and medial prefrontal, superior temporal, and lateral parietal regions, suggesting neural correlates improved working memory under cognitive remediation.

4. Discussion

The predominant paradigm of modern psychiatry posits that advances in neurosciences can unravel the mysteries of mental illness. Since the 1990s were declared the decade of the brain, imaging evidence has taught us a great deal about neural correlates of symptoms expression and recovery from an insult to the brain [79]. Despite remarkable neuroscientific advances, specific mechanisms behind major mental illnesses, thus far, have not been identified [80]. Moreover, whilst neurotransmitters are known to mediate synaptic pathways, research has not yet been able to explain any psychiatric disorder in terms of chemical imbalances [81]. Various reasons exist as to why neuroscience is unlikely to provide a definite understanding of the disordered mind. First and foremost, what is preventing the scientific strategy to reduce psychiatry to neuroscience is the fact that diagnoses listed in the Diagnostic and Statistical Manual of Mental Disorders' are not diseases but merely syndromes without a precise endophenotype [82]. Moreover, the pathways from temperamental vulnerabilities to illness cannot be understood without taking into account psychosocial adversities [83]. In this view, associations between biomarkers of pathological and treatment processes are unlikely to be strong or linear. Pharmacotherapy, whilst useful in severe mental disorders, it is not in any way curative, and psychosocial interventions continue to play an important role in psychiatric treatment, evoking multiple risk factors and complex interactive pathways to the disordered mind [84].

Research efforts in tandem with more powerful imaging techniques will further unravel the intricacy of cerebral organization behind pathological and treatment processes. Nonetheless, the scientific strategy to reduce psychiatry to neurosciences is hindered by a discrepancy between a clinical phenomenon and its neural substrate, which is rooted in a conceptual mind and brain gap.

5. Conclusion

Long before the era of functional neuroimaging it was suggested that intervention-driven changes in affect, cognition and behavior appear to have measurable biological analogues [85]. To date, the potential to characterize neural mechanisms of recovery processes have amassed vast neuroimaging data on treatment-induced brain plasticity. Pharmacotherapy and psychotherapy appear to engage neural circuits that are responsive to a discrete treatment modality. Although both have similar effects on brain activity patterns in patients who share the same diagnosis, their neural systems profile is not identical. While the former appears to act in a bottom- up manner on a subcortical level to regulate higher cortical structures, the latter acts top-down on cortical activity to subsequently impact subcortical regions. Although neuroimaging techniques have revolutionized our biological insight into recovery processes, little can be concluded about the precise neurobiological mechanism of these changes. The remaining question is whether these changes elucidate a neural mechanism of treatment action or simply reflect correlates of symptom amelioration. Despite methodological and theoretical limitations neuroimaging literature holds promise to strengthen the credibility and utility of mainstay in psychiatric treatment, and to improve clinical decision-making.

Conflict of interest

The author has no relevant affiliations or financial involvement with any organization or entity with a financial interest in or financial conflict with the subject matter or materials discussed in the manuscript. This includes employment, consultancies, honoraria, stock ownership or options, expert testimony, grants or patents received or pending, or royalties. No writing assistance was utilized in the production of this manuscript.

Author details

Maria Uscinska[1*], Andrea Polla Mattiot[2*] and Silvio Bellino[2*]

1 Centre for Personality Disorders, Department of Neurosciences, University of Turin, Italy

2 Centre for Personality Disorders, Molinette Hospital, Turin, Italy

*Address all correspondence to: maria.uscinska@edu.unito.it, andrea.polla.mattiot@gmail.com and silvio.bellino@unito.it

References

[1] Dougherty DD, Rauch SL, Rosenbaum JF. Essentials of Neuroimaging for Clinical Practice. Washington: American Psychiatric Association; 2004

[2] Fridriksson J, Morrow-Odom L, Moser D, Fridriksson A, Baylis G. Neural recruitment associated with anomia treatment in aphasia. NeuroImage. 2006;**32**(3):1403-1412

[3] Vetulani J, Sulser F. Action of various antidepressant treatments reduces reactivity of noradrenergic cyclic AMP-generating system in limbic forebrain. Nature. 1975;**257**(5526):495-496. DOI: 10.1038/257495a0

[4] Cleare A, Pariante CM, Young AH. Evidence-based guidelinesfor treating depressive disorders with antidepressants: A revision of the 2008 British Association for Psychopharmacology guidelines. Journal of Psychopharmacology. 2015;**29**:459-525

[5] Bauer M, Bschor T, Pfennig A. For the WFSBP task force on unipolar depressive disorders. World Federation of Societies of Biological Psychiatry (WFSBP) guidelines for biological treatment of unipolar depressive disorders in primary care. The World Journal of Biological Psychiatry. 2007;**8**:67-104

[6] Cipriani A, Furukawa TA, Salanti G. Comparative efficacy and acceptability of 12 newgeneration antidepressants: A multiple-treatments meta-analysis. Lancet. 2009;**373**:746-758

[7] Ross SB, Renyi AL. Inhibition of the uptake of tritiated 5-hydroxytryptamine in brain tissue. European Journal of Pharmacology. 1969;**7**:270-277

[8] Beck AT. The evolution of the cognitive model of depression and its neurobiological correlates. American Journal of Psychiatry. 2008;**165**:969-977

[9] Phillips ML, Drevets WC, Rauch SL, Lane R. Neurobiology of emotion perception II: Implications for major psychiatric disorders. Biological Psychiatry. 2003;**54**:515-528

[10] Ma Y. Neuropsychological mechanism underlying antidepressant effect: A systematic metaanalysis. Molecular Psychiatry. 2015;**20**:311-319

[11] Harmer CJ, Shelley NC, Cowen PJ, Goodwin GM. Increased positive versus affective perception and memory in healthy volunteers following selective serotonin and norepinephrine reuptake inhibition. The American Journal of Psychiatry. 2004;**161**:1256-1263

[12] Norbury R, Mackay CE, Cowen PJ, Goodwin GM, Harmer CJ. Short-term antidepressant treatment and facial processing, Functional magnetic resonance imaging study. The British Journal of Psychiatry. 2007;**90**:531-532

[13] Godlewska BR, Norbury R, Selvaraj S, Cowen PJ, Harmer CJ. Short-term SSRI treatment normalises amygdala hyperactivity in depressed patients. Psychological Medicine. 2012;**42**:2609-2617

[14] Shiroma PR, Thuras P, Johns B, Lim KO. Emotion recognition processing as early predictor of response to 8-week citalopram treatment in late-life depression. International Journal of Geriatric Psychiatry. 2014;**29**:1132-1139

[15] Leucht S, Komossa K, Rummel-Kluge C, et al. A meta-analysis of head to head comparisons of second-generation antipsychotics in the treatment of schizophrenia. The American Journal of Psychiatry. 2009;**166**:152-163

[16] Kendall T. The rise and fall of the atypical antipsychotics. The British Journal of Psychiatry. 2011;**199**:266-268

[17] Kapur S, Zipursky R, Jones C, Remington G, Houle S. Relationship between dopamine D_2 occupancy, clinical response, and side effects: A double-blind PET study of first-episode schizophrenia. The American Journal of Psychiatry. 2000;**157**:514-520

[18] Minzenberg MJ, Laird AR, Thelen S, Carter CS, Glahn DC. Meta-analysis of 41 functional neuroimaging studies of executive function in schizophrenia. Archives of General Psychiatry. 2009;**66**(8):811-822

[19] Lesh TA, Niendam TA, Minzenberg MJ, Carter CS. Cognitive control deficits in schizophrenia: Mechanisms and meaning. Neuropsychopharmacology. 2011;**36**(1):316-338

[20] Lesh TA, Tanase C, Geib BR, et al. A multimodal analysis of antipsychotic effects on brain structure and function in first-episode schizophrenia. JAMA Psychiatry. 2015;**72**:226-234

[21] Meisenzahl EM, Scheuerecker J, Zipse M, et al. Effects of treatment with the atypical neuroleptic quetiapine on working memory function: A functional MRI follow-up investigation. European Archives of Psychiatry and Clinical Neuroscience. 2006;**256**(8):522-531

[22] Jones HM, Brammer MJ, O'Toole M, et al. Cortical effects of quetiapine in first-episode schizophrenia: A preliminary functional magnetic resonance imaging study. Biological Psychiatry. 2004;**56**(12):938-942

[23] Honey GD, Bullmore ET, Soni W, Varatheesan M, Williams SC, Sharma T. Differences in frontal cortical activation by a working memory task after substitution of risperidone for typical antipsychotic drugs in patients with schizophrenia. Proceedings of the National Academy of Sciences of the United States of America. 1999;**96**(23):13432-13437

[24] Miller BJ, Buckley P, Seabolt W, Mellor A, Kirkpatrick B. Meta-analysis of cytokine alterations in schizophrenia: Clinical status and antipsychotic effects. Biological Psychiatry. 2011;**70**(7):663-671

[25] van Berckel BN, Bossong MG, Boellaard R, et al. Microglia activation in recent-onset schizophrenia: A quantitative (R)-[11C]PK11195 positron emission tomography study. Biological Psychiatry. 2008;**64**(9):820-822

[26] Pasternak O, Westin CF, Bouix S, et al. Excessive extracellular volume reveals a neurodegenerative pattern in schizophrenia onset. The Journal of Neuroscience. 2012;**32**(48):17365-17372

[27] Goodwin GM. Evidence-based guidelines for treating bipolar disorder: Revised second edition—recommendations from the British Association for Psychopharmacology. Journal of Psychopharmacology. 2009;**23**:346-388

[28] Yatham LN, Kennedy SH, Schaffer A, Parikh SV, Beaulieu S, O'Donovan C, et al. Canadian Network for Mood and Anxiety Treatments (CANMAT) and International Society for Bipolar Disorders (ISBD) collaborative update of CANMAT guidelines for the management of patients with bipolar disorder: Update 2009. Bipolar Disorders. 2009;**11**:225-255

[29] Grunze H, Vieta E, Goodwin GM, Bowden C, Licht RW, Möller HJ, et al. The World Federation of Societies of Biological Psychiatry (WFSBP) Guidelines for the Biological Treatment of Bipolar Disorders: Update 2010 on the treatment of acute bipolar depression. The World Journal of Biological Psychiatry. 2010;**11**:81-109

[30] Chisholm D, van Ommeren M, Ayuso-Mateos JL, Saxena S. Cost-effectiveness of clinical interventions for reducing the global burden of bipolar disorder. The British Journal of Psychiatry. 2005;**187**:559-567

[31] Green MJ, Cahill CM, Malhi GS. The cognitive and neurophysiological basis of emotion dysregulation in bipolar disorder. Journal of Affective Disorders. 2007;**103**:29-42

[32] Strakowski SM, Adler CM, Holland SK, Mills NP, DelBello MP, Eliassen JC. Abnormal fMRI brain activation in euthymic bipolar disorder patients during a counting Stroop interference task. The American Journal of Psychiatry. 2005;**162**:1697-1705

[33] Gabrieli S, Makris N, Laviolette P. An fMRI study of working memory in persons with bipolar disorder or at genetic risk for bipolar disorder. American Journal of Medical Genetics. Part B, Neuropsychiatric Genetics. 2010;**153B**:120-131

[34] Leibenluft E, Rich BA, Vinton DT, Nelson EE, Fromm SJ, Berghorst LH, et al. Neural circuitry engaged during unsuccessful motor inhibition in pediatric bipolar disorder. The American Journal of Psychiatry. 2007;**164**:52-60

[35] Rich BA, Vinton DT, Roberson-Nay R, Hommer RE, Berghorst LH, McClure EB, et al. Limbic hyper-activation during processing of neutral facial expressions in children with bipolar disorder. Proceedings of the National Academy of Sciences of the United States of America. 2006;**103**:8900-8905

[36] Kronhaus DM, Lawrence NS, Williams AM, Frangou S, Brammer MJ, Williams SCR, et al. Stroop performance in bipolar disorder: Further evidence for abnormalities in the ventral prefrontal cortex. Bipolar Disorders. 2006;**8**:28-39

[37] Altshuler LL, Bookheimer SY, Townsend J, Proenza MA, Eisenberger N, Sabb F, et al. Blunted activation in orbitofrontal cortex during mania: A functional magnetic resonance imaging study. Biological Psychiatry. 2005;**58**:763-769

[38] Lawrence NS, Williams AM, Surguladze S, Giampietro V, Brammer MJ, Andrew C, et al. Subcortical and ventral prefrontal cortical neural responses to facial expressions distinguish patients with bipolar disorder and major depression. Biological Psychiatry. 2004;**55**:578-587

[39] Gruber SA, Rogowska J, Yurgelun-Todd DA. Decreased activation of the anterior cingulate in bipolar patients: An fMRI study. Journal of Affective Disorders. 2004;**82**:191-201

[40] Adler CM, Holland SK, Schmithorst V, Tuchfarber MJ, Strakowski SM. Changes in neuronal activation in patients with bipolar disorder during performance of a working memory task. Bipolar Disorders. 2004;**6**:540-549

[41] Blumberg HP, Leung HC, Skudlarski P, Lacadie CM, Fredericks CA, Harris BC, et al. A functional magnetic resonance imaging study of bipolar disorder: State- and trait-related dysfunction in ventral prefrontal cortices. Archives of General Psychiatry. 2003;**60**:601-609

[42] Blumberg HP, Donegan NH, Sanislow CA, Collins S, Lacadie C, Skudlarski P, et al. Preliminary evidence for medication effects on functional abnormalities in the amygdala and anterior cingulate in bipolar disorder. Psychopharmacology. 2005;**183**:308-313

[43] Yurgelun-Todd DA, Gruber SA, Kanayama G, Killgore WD, Baird AA, Young AD. fMRI during affect discrimination in bipolar affective disorder. Bipolar Disorders. 2000;**2**:237-248

[44] Kandel ER. A new intellectual framework for psychiatry. American Journal of Psychiatry. 1998;**155**:457-469

[45] Beutel ME, Stern E, Silbersweig DA. The emerging dialogue between psychoanalysis and neuroscience: Neuroimaging perspectives. Journal of the American Psychoanalytic Association. 2003;**51**:773-801

[46] Ochsner K, Bunge SA, Gross JJ, Gabrieli JDE. Rethinking feelings: An fMRI study of the cognitive regulation of emotion. Journal of Cognitive Neuroscience. 2012;**14**:1215-1229

[47] Goldapple K, Segal Z, Garson C, Lau M, Bieling P, Kennedy S, et al. Modulation of cortical-limbic pathways in major depression: Treatment specific effects of cognitive behav ioral therapy. Archives of General Psychiatry. 2004;**61**:34-41

[48] Seminowicz DA, Mayberg HS, McIntosh AR, Goldapple KK, Kennedy S, Segal Z, et al. Limbic-frontal circuitry in major depression: A path modeling metanalysis. NeuroImage. 2004;**22**:409-418

[49] Paquette V, Levesque J, Mensour B, Leroux JM, Beaudoin G, Bourgouin P, et al. 'Change the mind and you change the brain': Effects of cognitive-behavioral therapy on the neural correlates of spider phobia. NeuroImage. 2003;**18**:401-409

[50] Oldham JM. Borderline personality disorder and suicidality. The American Journal of Psychiatry. 2006;**163**(1):20-26

[51] Linehan MM, Comtois KA, Murray AM. Two year randomised controlled trial and follow-up versus treatment by experts for suicidal behaviours and borderline personality disorder. Archives of General Psychiatry. 2006;**63**(7):757-766

[52] Lynch WT, Trost N, Salsman N, Linehan MM. Dialectical behavior therapy for borderline personality disorder. Annual Review of Clinical Psychology. 2007;**3**(1):181-205

[53] Schmitt R, Winter D, Niedtfeld I, Herpertz SC, Schmahl C. Effects of psychotherapy on neuronal correlates of reappraisal in female patients with borderline personality disorder. Biological Psychiatry: Cognitive Neuroscience and Neuroimaging. 2016;**1**(6):548-557

[54] Winter D, Niedtfeld I, Schmitt R, Bohus M, Schmahl C, Herpertz SC. Neural correlates of distraction in borderline personality disorder before and after dialectical behavior therapy. European Archives of Psychiatry and Clinical Neuroscience. 2016;**267**(1):51-62

[55] Schulze L, Schmahl C, Niedtfeld I. Neural correlates of disturbed emotion processing in borderline personality disorder: A multimodal meta-analysis. Biological Psychiatry. 2016;**79**(2):97-106

[56] Martin SD, Martin RMN, Rai SS, Richardson MA, Royall R, Eng IEE. Brain blood flow changes in depressed patients treated with interpersonal psychotherapy or venlafaxine hydrochloride: Preliminary findings. Archives of General Psychiatry. 2001;**58**:641-648

[57] Brody AL, Saxena S, Stoessel P, Gillies LA, Fairbanks LA, Alborzian S, et al. Regional brain metabolic changes in patients with major depression treated with either paroxetine or interpersonal therapy: Preliminary findings. Archives of General Psychiatry. 2001a;**58**:631-640

[58] Brody AL, Saxena S, Mandelkern MA, Fairbanks LA, Ho ML, Baxter LR. Brain metabolic changes associated with symptom factor improvement in major depressive disorder. Biological Psychiatry. 2001b;**50**:171-178

[59] Tohen M, Hennen J, Zarate CM, Baldessarini RJ, Strakowski SM, Stoll AL, et al. Two-year syndromal and functional recovery in 219 cases of first-episode major affective disorder with psychotic features. American Journal of Psychiatry. 2000;**157**:220-228

[60] Swartz HA, Frank E. Psychotherapy for bipolar depression: A phase-specific treatment strategy? Bipolar Disorders. 2001;**3**:11-22

[61] Zaretsky A. Targeted psychosocial interventions for bipolar disorder. Bipolar Disorders. 2003;**5**:80-87

[62] Colom F, Vieta E. A perspective on the use of psychoeducation, cognitive behavioral therapy and interpersonal therapy for bipolar patients. Bipolar Disorders. 2004;**6**:480-486

[63] Miklowitz DJ. Adjunctive psychotherapy for bipolar disorder: State of the evidence. The American Journal of Psychiatry. 2008;**165**:1408

[64] Perry A, Tarrier N, Morriss R, McCarthy E, Limb K. Randomised controlled trial of efficacy of teaching patients with bipolar disorder to identify early symptoms of relapse and obtain treatment. British Medical Journal. 1999;**318**:149-153

[65] Colom F, Vieta E, Reinares M, Martínez-Arán A, Torrent C, Goikolea JM, et al. Psychoeducation efficacy in bipolar disorders: Beyond compliance enhancement. Journal of Clinical Psychiatry. 2003b;**64**:1101

[66] Favre S, Aubry JM, Gex-Fabry M, Ragama-Pardos E, McQuillan A, Bertschy G. Translation and validation of a french version of the young mania rating scale. L'Encéphale. 2003;**29**:499

[67] Grynszpan O, Perbal S, Pelissolo A, Fossati P, Jouvent R, Dubal S, et al. 2011. Efficacy and specificity of computer-assisted cognitive remediation in schizophrenia: A meta-analytical study. Psychological Medicine. 2011;**41**:163-173

[68] Naismith SL, Redoblado-Hodge MA, Lewis SJG, Scott EM, Hickie IB. Cognitive training in affective disorders improves memory: A preliminary study using the NEAR approach. Journal of Affective Disorders. 2010;**121**:258-262

[69] McGurk SR, Twamley EW, Sitzer DI, McHugo GJ, Mueser KT. A metaanalysis of cognitive remediation in schizophrenia. American Journal of Psychiatry. 2007;**164**:1791-1802

[70] Penadés R, Pujol N, Catalán R, Massana G, Rametti G, García-Rizo C, et al. Brain effects of cognitive remediation therapy in schizophrenia: A structural and functional neuroimaging study. Biological Psychiatry. 2013;**73**(10): 1015-1023

[71] Bor J, Brunelin J, d'Amato T, Costes N, Suaud-Chagny MF, Saoud M, et al. How can cognitive remediation therapy modulate brain activations in schizophrenia? An fMRI study. Psychiatry Research. 2011;**192**(3):160-166. DOI: 10.1016/j. pscychresns. 2010. 12. 004

[72] Wexler BE, Anderson M, Fulbright RK, Gore J. Preliminary evidence of improved verbal working memory performance and normalization of taskrelated frontal lobe activation in schizophrenia following cognitive exercises. The American Journal of Psychiatry. 2000;**157**:1694-1697

[73] Meusel LA, Hall GB, Fougere P, McKinnon MC, MacQueen GM. Neural correlates of cognitive remediation in patients with mood disorders. Psychiatry Research. 2013;**214**(2):142-152. DOI: 10.1016/j. pscychresns.2013.06.007

[74] Rosen AC, Sugiura L, Kramer JH, Whitfield-Gabrieli S, Gabrieli JD. Cognitive training changes hippocampal function in mild cognitive impairment:

A pilot study. Journal of Alzheimer's Disease. 2011;**26**(Suppl. 3):349-357

[75] Schweizer S, Grahn J, Hampshire A, Mobbs D, Dalgleish T. Training the emotional brain: Improving affective control through emotional working memory training. Journal of Neuroscience. 2013;**33**:5301-5311

[76] Thorsen A, Johansson K, Løberg EM. Neurobiology of cognitive remediation therapy for schizophrenia: A systematic review. Frontiers in Psychiatry. 2014;**5**:103. DOI: 10.3389/fpsyt.2014.00103

[77] Isaac C, Januel D. Neural correlates of cognitive improvements following cognitive remediation in schizophrenia: A systematic review of randomized trials. Socioaffective Neuroscience & Psychology. 2016;**6**:30054

[78] Kurtz MM, Gerraty R. A meta-analytic investigation of neurocognitive deficits in bipolar illness: Profile and effects of clinical state. Neuropsychology. 2009;**23**:551-562

[79] Compston A. Decade of the brain. Brain. 2005;**128**:1741-1742

[80] Nemeroff CB, Kilts CD, Berns GS. Functional brain imaging: Twenty-first century phrenology or psychobiological advance for the millennium? The American Journal of Psychiatry. 1999;**15**:671-673

[81] Valenstein E. Blaming the Brain: The Truth About Drugs and Mental Health. New York (NY): Free Press; 1998

[82] Kendler KS, Prescott CA. Genes, Environment, and Psychopathology: Understanding the Causes of Psychiatric and Substance Use Disorders. New York (NY): Guilford Press; 2006

[83] Caspi A, Sugden K, Moffitt TE, et al. Influence of life stress on depression: Moderation by a polymorphism in the 5-HTT gene. Science. 2003;**301**:386-389

[84] Paris J. Prescriptions for the Mind: A Critical View of Contemporary Psychiatry. New York (NY): Oxford University Press; 2008

[85] Freud S. Project for a Scientific Psychology. The Standard Edition of the Complete Psychological Works of Sigmund Freud. London: Hogarth Press; 1895. pp. 283-397

Chapter 8

Aberrant Brain Neuroplasticity and Function in Drug Addiction: A Focus on Learning-Related Brain Regions

Patricia Sampedro-Piquero, Luis J. Santín and Estela Castilla-Ortega

Abstract

This chapter will review the altered brain structure and function associated to drug addiction, with a focus on brain regions involved in learning and motivated behavior. As evidenced by both clinical and preclinical studies, repeated drug exposure affects whole brain neuroplasticity including the mesolimbic system which is a main locus for reward, an action-control center such as the dorsal striatum, and limbic brain regions such as the prefrontal cortex, the hippocampus, and the amygdala that are involved in behavioral control, memory, and mood. In this way, the drug-seeking actions that were initially intentional responses become involuntary habits governed by the dorsal striatum. Drug addiction may also curse with a reduced ability to experience rewards that are unrelated to drugs and emotional dysregulation, while the impairment on limbic regions contributes to generate cognitive symptoms. These entail persistent memories for previous experiences with the drug contrasting with a global cognitive decline that may hamper the acquisition of new, adaptive learnings. Overall, these features promote a desire for the drug, leading to relapse in drug use. Further drug exposure, in turn, aggravates its consequences on the brain and behavior, creating the harmful "addiction cycle."

Keywords: substance use disorders, habits, motivation, memory, mood, accumbens, striatum, limbic regions

1. Introduction

The use of psychoactive drugs that induce dependence (including psychostimulants (such as cocaine, methamphetamine, etc.), opioids (heroin, methadone, etc.), cannabinoids, tobacco, and alcohol, among others) is widely extended in the first world countries [1, 2]. The widespread drug use entails a main socioeconomic burden, because drug use is associated to antisocial behavior and delinquency, violence and accidents, social exclusion, physical and psychiatric illnesses, and even disability and death [1, 2]. In this regard, it is worth mentioning that a recent global study identified alcohol as the leading risk factor for premature death in the population aged 15–49 years [1]. Considering the severity of the drug use problem, the World

Health Organization currently destines efforts for substance abuse management in order to improve both treatment and prevention programs (https://www.who.int/substance_abuse/publications/drugs/en/).

Nevertheless, while drug use entails significant risks, regular usage of drugs is not a synonym of suffering a drug addiction disorder. Drug addiction (or substance use disorder—SUD) is a chronic disorder with a high relapse rate, in which the person "loses control" over drug intake despite the negative consequences on their daily life and even against the desire to remain abstinent [3]. Drug addiction may only be experienced by a subgroup of more "vulnerable" individuals that get in contact with drugs. Specifically, approximately 11% of people that use drugs would develop a SUD, meaning an uncontrollable and harmful drug use pattern that may need treatment [2]. Therefore, the scientific community has invested in investigating those factors or mechanisms that cause and explain the onset and maintenance of a SUD. As the deleterious impact of addictive drugs on the brain—the organ that controls behavior—became evident, addiction has been considered as a "brain disease" [4]. The current "brain disease" model of addiction has important implications for SUD prevention and treatment, since medical interventions that regulate brain functioning (e.g., pharmacotherapy) may be valid for addiction, and persons with SUDs may benefit for public treatment policies reserved to other medical illnesses, while the social stigma is attenuated since drug addiction is a medical condition instead of a voluntary choice or an hedonistic act [4]. However, this model is not exempt of criticism [5, 6], partially because the relevance of social and psychological factors is diminished in favor of the biological elements, and freeing the person from responsibility underestimates the importance of the personal willpower and motivation toward therapeutic change.

Setting this controversy aside, there is a consensus in that drug addiction, being a "brain disease" or not, certainly involves a neurobiological brain dysfunction that affects behavior. Brain morphological alterations in persons using different drug types (such as alcohol, cannabis, cocaine, methamphetamine, heroin, or tobacco) have been consistently reported even at the macrostructural level, usually involving significant gray and/or white matter shrinkage [7–13]. Moreover, functional neuroimage techniques reveal that connectivity among brain regions is also dysregulated [14]. It is important to note that the aberrant brain structure and function associated to drug addiction most likely results from a combination of (biological) brain features that exist *previous to* drug use as vulnerability factors, with the neuroadaptations that are *induced by* the drug itself (**Figure 1A**). Solid evidence has been provided in both ways (reviewed in [15]). On the one hand, individual differences in the form of stable personality traits such as impulsivity, elevated anxiety, risk-taking, and sensation seeking that are assumed to entail a particular biological and brain basis [16, 17] may predispose to engage in both drug use and addiction. On the other hand, brain and behavioral abnormalities often correlate with drug use patterns (i.e., the amount of drug consumed and/or the number of years using the drug) and may be completely or partially recovered by protracted drug abstinence [7, 8, 11, 13, 15, 18], suggesting that they were directly induced by the continuous action of the drug. Notably, preclinical studies in laboratory animals (that allow the exposure to the drug to be controlled by the experimenter) have confirmed both evidences. Individual traits in rodents (e.g., increased impulsivity) predict their subsequently exacerbated response to drugs compared to rodents that do not show this feature (e.g., [19, 20]); and both brain and behavioral alterations are experimentally induced by administering drugs to naïve animals (e.g., [21–23]).

Therefore, while it is difficult—especially for clinical research—to elucidate whether the observed behavioral and brain features are cause or consequence of drug use, both drug vulnerability factors and drug-induced brain effects are likely

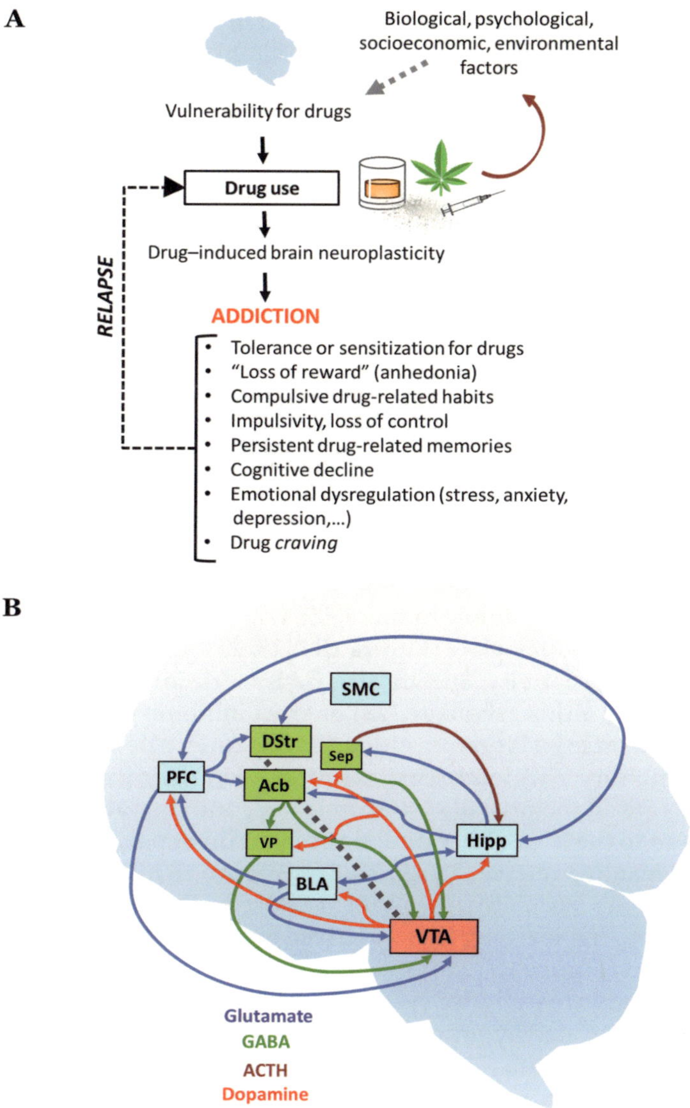

Figure 1.
(A) The "drug addiction cycle." Numerous factors intervene in the vulnerability for drugs, including a "vulnerable brain." Drug consumption induces widespread brain neuroadaptations that, in vulnerable individuals, would be addiction-like behavioral alterations that are likely to promote further drug use, aggravating its effects. (B) A non-exhaustive schematic representation of the brain structures and connections involved the brain circuit of learning, reward, and motivated behavior. A maladaptive functioning of this circuit supports the etiology and maintenance of drug addiction. Brain structures are colored on the basis of their main neurochemical content. The dashed line represents the "spiraling" nigrostriatal connections. Abbreviations: Acb, accumbens; ACTH, acetylcholine; BLA, basolateral amygdala; Dstr, dorsal striatum; GABA, γ-aminobutyric acid; Hipp, hippocampus; PFC, prefrontal cortex; Sep, septum; SMC, sensorimotor cortex; VP, ventral pallidum; VTA, ventral tegmental area.

to coexist and be interrelated. In the worst case scenario, a "vulnerable" brain is exposed to the drug, triggering an exacerbated response to the substance that increases the amount of drug subsequently consumed, thus also increasing the potential drug-induced harm (**Figure 1A**). Without the intent of underestimating the notable importance of psychological, social, economic, and environmental factors in the etiology and maintenance of drug addiction, this chapter will focus on the neurobiological component. In particular, we will review that the integrity of key brain regions that are normally involved in control of reward, planning, learning, and motivated behavior is compromised in drug addiction to favor uncontrollable drug intake as well as other behavioral symptoms. Specifically, we will focus on the mesolimbic system, the dorsal striatum, and the limbic regions as key components of the "brain addiction circuit."

2. The mesolimbic system: a locus for drug and non-drug-related rewards

2.1 Experiencing rewards and learning to predict them

The mesolimbic system has been a traditional focus of drug addiction research, since it is a key substrate for reward and motivated behavior. The mesolimbic system comprises the accumbens (also called "ventral striatum") and the midbrain ventral tegmental area (VTA) as its main brain nodes and dopamine (often considered as the molecule of "pleasure and happiness" [24]) as its major neurotransmitter [25, 26]. The dopaminergic projection neurons in the VTA release dopamine to the accumbens—either at its core or shell subdivisions—as well as to memory-related limbic brain regions such as the prefrontal cortex, the hippocampus, and the amygdala (**Figure 1B**) [25, 27]. Conversely, these brain regions regulate VTA activity. Specifically, GABAergic inhibitory pathways from the accumbens may either stimulate [28] or exert inhibitory feedback control [29] over dopamine release by targeting either the dopaminergic VTA projection neurons or the inhibitory VTA interneurons [30, 31]. For their part, the glutamatergic limbic regions are all reciprocally interconnected, and they also project to the accumbens and to the VTA either directly or by indirect polysynaptic pathways, to stimulate dopamine release [27, 28, 32, 33] (**Figure 1B**). This illustrates that reward and memory systems in the brain are closely interrelated, which makes sense considering that learning is often driven by rewards, punishments, and their anticipation (**Figure 1B**) [34].

The dopaminergic mesolimbic system is involved in experiencing pleasure, and it is directly activated by primary rewards such as palatable food or sexual behavior [24], novel stimuli [35], or pleasant music [36]. By engaging its reciprocal connections to the limbic regions, the accumbens is important for determining the motivational valence of stimuli and for assessing learning incentives. In other words, the accumbens discriminates appetitive from aversive stimuli and decides in which degree they are "liked" or "wanted" [24, 37]. In agreement to this, preclinical research reveals a role of the accumbens in many forms of learning such as in spatial navigation [38], novel object and place recognition [39], fear conditioning [40], or instrumental behavior [41] (see "preclinical models of learning" in **Box 1**), and dopamine in the mesolimbic system promotes an activated state of alertness, arousal, or "seeking" that would facilitate exploration and reward gathering [42]. Moreover, the accumbens has an important role in anticipating the occurrence of rewards by learning which stimuli predicts them (i.e., acquiring conditioned reward-stimuli associations; **Table 1**) [43]. By association with a rewarding stimulus, a neutral stimulus becomes a conditioned reward and gains incentive motivational salience, being able to activate the mesolimbic reward system by itself [34].

When in the presence of dependence-inducing drugs, the dopaminergic mesolimbic system is highly activated, engaging different neurobiological mechanisms depending on the substance (e.g., inhibition of dopamine reuptake cocaine and methamphetamine [44, 45], stimulation of dopaminergic VTA neurons alcohol, methamphetamine, nicotine, cannabinoids [46–49], inhibition

Anhedonia: A reduced ability to feel pleasure or joy; a loss of interest for activities or stimuli that were previously engaging for the individual and elicited positive emotions. It is often a symptom of low mood (depression-like behavior). Persons with SUDs may suffer anhedonia or "loss of reward" for experiences that are not related to drugs.

Appetitive, aversive: Qualities of stimuli: rewarding (appetitive) or disliking (aversive).

Craving: An intense, uncontrollable, and anxious desire to use the drug. It is usually elicited by drug-associated stimuli and it may lead to relapse in drug use.

Declarative memory: This memory overlaps with the most common concept of "memory" as it refers to the ability to learn (and also recall, forget, etc.) facts, concepts, or words, life events, and spatial or contextual stimuli (e.g., when America was discovered, what you had for dinner yesterday, where the car was stationed, etc.).

Dorsal striatum: A motor control brain center that works in consonance with cortical brain regions (cortico-striatal circuit) to select and initiate appropriate goal-directed responses. The dorsal striatum also transforms the goal-directed actions that are repeatedly rewarded into automatic habits.

Drug sensitization (vs drug tolerance): Exacerbation of the rewarding or psychomotor effects of the drug, as a result of the neuroadaptations induced by repeated drug exposure. There is also evidence of the opposite effect, drug tolerance, meaning that the drug progressively blunts its actions.

Drug-associated stimuli: Those stimuli (objects, places, people, feelings, etc.) that, by associative learning processes, have been "linked" to the effects of the drug or to drug availability. The presence of these stimuli is a main cause of relapse, as they trigger both craving feelings and uncontrollable drug-seeking and drug-taking habits.

Escalation (in drug intake): The phenomenon by which the person progressively increases drug use, leading to excessive drug intake. It is also evidenced in the preclinical drug self-administration model, where the animal progressively self-administers more quantities of the drug as the task progresses.

Executive functions: A set of high-level cognitive skills that is important for "ruling" behavior. They involve decision-making, planning, reasoning, attentional control, cognitive flexibility, inhibition of undesired behaviors, etc.

Goal-directed behavior: Response directed to obtain a reward. It is planned, conscious, and often useful.

Habits: "Automatic" and involuntary responses that require minimal cognitive resources to be executed. They are generated after a goal-directed response has been repeated and rewarded numerous times. While habits are adaptive for everyday functioning, a main problem in drug addiction is that behaviors associated to drugs (drug-seeking, drug-taking, etc.) also become uncontrollable habits, contributing to relapse in drug use.

Incentive (motivational) salience: Refers to the intensity of attention, attraction, or desire ("wanting") that is elicited by a stimulus. It is usually related to its rewarding value. Drugs and drug-related stimuli gain incentive motivational salience in addiction.

Limbic regions: Brain regions mainly involved in the regulation of cognition and emotion. This review considers the prefrontal cortex, the hippocampus, and the amygdala as main brain limbic areas. They are impaired by addictive drugs.

Long-term potentiation (LTP), long-term depression (LTD): A form of neuroplasticity that changes the strength of a synapse, for example, as a result of learning processes or after exposure to a drug of abuse. In the LTP, the postsynaptic neuron increases its response (e.g., more neurotransmitter is released, or more neurotransmitter receptors are generated), while in the LTD the postsynaptic response is debilitated.

Mesolimbic system: Brain system mainly comprised by the VTA and the accumbens. It is important for experiencing, predicting, and assessing rewards and thus for motivated (i.e., goal-directed) behavior. It is also involved in the motor-activating effects of drugs.

Neuron: The main nerve cell in the brain that processes and transmits information through the synapsis. The main parts of a neuron are depicted in **Figure 2**. **Projection neurons** possess long axons that allow communication between distant brain regions, while **interneurons** have shorter axons, limited to a single brain area.

Neuroplasticity: Neuroplasticity or neuroadaptation refers to changes in the anatomical structure (dendrites, axon, nuclei, etc.) and function (synaptic strength, neurotransmitter release, etc.) of neurons, in response to environmental or internal stimuli. Another form of neuroplasticity is the generation of new neurons in the adult brain (adult hippocampal neurogenesis). Brain neuroplasticity is modulated by drugs of abuse, yielding an aberrant pattern of brain functioning that contributes to generate and maintain addiction.

Neurotoxicity: The effect of a hazardous substance that may involve an irreversible loss of the neuron's anatomy and function and even its death. Addictive drugs such as alcohol, methamphetamine, or heroin have demonstrated neurotoxicity.

Neurotransmitters: Chemical messengers synthetized by the neurons that transmit information between them, acting on specific receptors in the synapse. **Glutamate** is the main excitatory brain neurotransmitter, as it "activates" the target neuron, while **GABA** has an inhibitory role; **dopamine** is critical in the mesolimbic reward system regulating reward and arousal.

Preclinical models of addiction: Paradigms or "tasks" that are designed to assess addiction-related behaviors in animals, usually rodents (i.e., "preclinical" refers to laboratory research—animal or in vitro—previous to clinical studies in humans). For example, the **drug-induced conditioned place preference** assesses drug reward and the learning of drug-stimuli associations, by examining how much the rodent prefers to stay in a maze compartment where the drug was previously administered (which is distinguishable from a neutral maze compartment as they have different contextual cues). The **drug self-administration** paradigm assesses motivation for the drug, by examining how much the rodent presses a level that results in drug delivery (or how much the rodent keeps insisting in pressing the lever even when the drug is no longer provided). Pressing the lever is considered as a drug-seeking (or taking) behavior. There are other models such as the "voluntary drinking" paradigms used for ethanol.

Preclinical models of learning: There are a wide variety of tasks to assess different forms of learning and memory in rodents. For example, in the **spatial navigation** tasks, the animal learns to orientate in the surrounding space to find a particular place in a maze (e.g., the maze exit, or hidden food rewards). In **novelty-based** tasks, the animal prefers to explore a novel object or place as long as it remembers the familiar one(s). Tasks based on **associative learning** require the animal to associate stimuli; for example, in fear conditioning, a compartment of the maze where an electric shock is provided is discriminated by its contextual cues (shock-cue association). In **instrumental learning tasks** (operant conditioning), the animal learns to perform a specific response, such as pressing a lever, to obtain a reward, such as food.

Psychiatric comorbidity: Different psychiatric disorders that occur simultaneously in the same individual, usually worsening the therapeutic outcome. Drug addiction is often associated to high psychiatric comorbidity, including mood and anxiety disorders (depression, generalized anxiety, phobias, etc.) and personality disorders.

Relapse: Resuming drug use after a period of abstinence.

Reward: A stimuli or outcome that is pleasurable and/or beneficial for the individual. **Primary rewards** are those stimuli intrinsically pleasurable (e.g., a delicious food), while **conditioned rewards** are those that have gained their reinforcing value by being associated with a rewarding stimuli (e.g., the sound of a bell that rings when the food is ready to serve becomes rewarding).

Substance use disorder (SUD), drug addiction: A chronic and highly relapsing disorder which its main characteristic is an uncontrollable (and usually excessive) drug intake. Furthermore, addiction frequently carries a socioeconomic and health burden for the individual, including motivational, emotional, and cognitive impairment. Unfortunately, these symptoms induced by drugs contribute to further drug use, generating a "vicious cycle."

Stress: A physiological response generated by a stimulus perceived as threatening or aversive. The stress response is dysregulated by addictive drugs, and experiencing stress contributes to relapse in drug use.

Synapse: It is the region where chemical or electrical information is transmitted from one neuron to another. A typical chemical synapse uses neurotransmitters that are synthetized by the presynaptic neuron and released through the axon terminals (**Figure 2**); and then they bind with specific neurotransmitter receptors in the postsynaptic neuron that may be located in the dendritic spines but also in the axon or soma.

Vulnerability: Referred to addiction, "vulnerability" entails both biological features and behavioral (personality) traits that predispose the individual to initiate and maintain drug use or to generate a SUD. For example, exacerbated impulsivity and reduced inhibition of behavior, inclination to take risks, preference for experiencing novel stimuli or seeking sensations, or an anxious personality are associated to increased risk for drug abuse. Importantly, the behavioral attributes are assumed to be a reflection of a particular pattern of brain functioning (i.e., they entail a biological-brain-correlate).

Working memory: It is a short-term memory capacity for concepts or stimuli that do not need to be remembered in the long-term, but they should be processed (i.e., mentally manipulated or "worked with") for a short period of time. For example, working memory is needed for reasoning, planning, and solving problems or mathematical operations.

Box 1.
Definitions.

of VTA GABAergic interneurons opioids and cocaine [50, 51]. According to the accumbens' role for experiencing rewards, the accumbens is involved in enjoying the recreational feelings induced by drugs [52], in their "activating" psychomotor effects [53, 54], and in learning the stimuli that are predictive of the drug's effects or its availability (i.e., drug-stimuli associations [43, 55]). In this way, rodents with lesions in the accumbens will reduce the expression of drug-seeking or drug-taking behaviors when they are tested in common preclinical models for addiction-like responses, such as conditioned place preference or self-administration paradigms [54, 56–58].

	"Normal" functioning	Drug addiction
Mesolimbic system	• Reward • Incentive assessment ("liking") • Associative learning, reward prediction • Alertness, activation	• Drug reward (sensitization, tolerance) • Anhedonia • Incentive assessment ("wanting" drugs) • Drug-stimuli associations • Drug psychomotor activation • Drug craving • Relapse
Dorsal striatum	• Motor learning • Behavior-optimizing habits • Goal-directed responses	• Drug-related habits • Drug craving • Relapse
Limbic regions	• Behavioral control • Executive functions • Working memory • Declarative memory (associative learning) • Emotional regulation	• Behavioral disinhibition • Cognitive decline • Drug-stimuli associations • Emotional dysregulation • Drug craving • Relapse

*Behavioral functions and addiction symptoms are linked to the main brain region(s) that supports them, but it should be noted that these reward and learning-related brain systems act in close synchrony (**Figure 1B**) to support behavior.*

Table 1.
Impaired function of learning-related brain regions in drug addiction.

2.2 Desire for drug overcomes natural rewards in addiction

As a result of its repeated activation by chronic drug exposure, the mesolimbic system may undergo long-lasting neuroadaptations which are involved in addiction (**Figure 2**). In clinical population with SUDs, a reduced volume of the accumbens has been reported [59, 60], and one *postmortem* study in cocaine users reveals a loss of dopaminergic neurons in the midbrain [61]. In drug-withdrawn animals, experiments have described persistent changes in accumbens dendrite branching and spine density (that are normally increased for alcohol, cocaine, methamphetamine, and nicotine [62, 63] but decreased for morphine or cannabinoids [63, 64]) as well as in the VTA (where psychostimulants tend to increase dendritic arborization and spines [65] but cannabinoids and opioids induce visible morphometrical reductions in the soma of the dopaminergic neurons [64, 66]).

Importantly, these structural modifications concomitantly occur with profound neurochemical and functional changes (**Figure 2**, **Table 1**), including modifications of the synaptic strength (long-term potentiation, LTP, or long-term depression, LTD) [67, 68]. The drug-induced neuroplastic and neurochemical adaptations, involving dopamine and glutamate signaling [69, 70], may augment the mesolimbic response to the drug. This supports the phenomenon of "behavioral sensitization," referring to an exacerbated drug's rewarding or motor-activating effects. Drug sensitization has been widely reported in rodents that will progressively increase locomotor activity and VTA dopamine release after they are repeatedly exposed to moderate doses of commonly abused drugs (most frequently to psychostimulants, but also to other drug types [70, 71]). But the evidence of drug sensitization in humans is more scarce [72]. In fact, there is evidence against

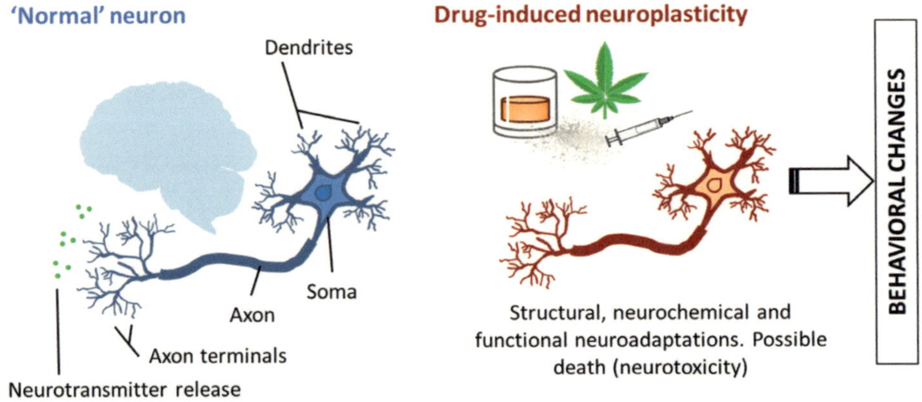

Mesolimbic system	Dorsal striatum	Limbic regions
Accumbens and VTA: • Changes in dendrite branching and spine density • Functional synaptic changes (LTP and/or LTD) **Accumbens:** • Reduced gray matter volume • Neurochemical alterations (lower levels of dopamine and D2 receptor expression; altered glutamate signalling) **VTA:** • Loss of dopaminergic VTA neurons (cocaine), soma reduction (cannabinoids, opioids) • Exhacerbated or decreased dopamine release	• Changes in dendrite branching and spine density • Functional synaptic changes (LTP and/or LTD) • Neurochemical alterations (lower levels of dopamine and D2 receptor expression; potentiated glutamate signalling)	**Prefrontal cortex, hippocampus, amygdala:** • Changes in dendrite branching and spine density • Functional synaptic changes (LTP and/or LTD) • Reduced gray matter volume • Neurochemical alterations (GABA and glutamate signalling) **Hippocampus:** • Reduced adult hippocampal neurogenesis

Figure 2.
A non-exhaustive list of brain structural and functional neuroadaptations induced by addictive drugs. Only the brain areas that are the focus of this review are depicted; but it should be noted that drugs affect widespread neuroplasticity in the whole brain. Neurotoxic effects (i.e., neuronal death) have been evidenced for certain drug types such as alcohol, methamphetamine, or heroin.

the drug-sensitization theory, reporting that the drug-induced dopamine response could become progressively blunted or habituated, which would then induce drug tolerance effects instead [72, 73]. Drug tolerance may ultimately lead to increased drug use, since more quantity of the substance is progressively needed to experience its effects.

In any case, escalation in drug intake is associated to a notable reduction of basal dopaminergic transmission in the accumbens and in the whole striatum, as evidenced by lower levels of endogenous striatal dopamine and reduced expression of dopamine receptors—mostly the postsynaptic D2 receptor [74–77]. This may contribute to the fact that, contrasting with the ability of drugs to stimulate the mesolimbic system, primary rewards may diminish their reinforcing value in addiction [73, 78]. Accordingly, an increased brain threshold for experiencing reward (measured by intracranial self-stimulation) and "loss of pleasure" anhedonic behaviors (e.g., reduced intake of a highly palatable food) are described in drug-withdrawn animals (reviewed in [79]). A diminished interest for non-drug rewards will impede persons with SUD to enjoy daily-life experiences or to attain interpersonal and professional goals as they now hold a weak appeal [72]. As predicted by this "loss of reward" model, drug use may then gain motivational incentive as a compensation for the decreased sensitivity to natural rewards and the hypodopaminergic mesolimbic state [73, 78, 79].

Thereby, while in a state of overall reduced reward and motivation for non-drug experiences, the drug and its associated stimuli would increase their incentive value in addiction: drugs would be "wanted," even when they are no longer "liked" [80]. In relation to this, drug use is highly driven by "craving", an intense and uncontrollable desire for the drug that progressively increases during abstinence periods (i.e., craving incubation) and is greatly triggered or aggravated when drug-associated stimuli are presented, eliciting relapse in compulsive drug-seeking or drug-taking [55]. Current evidence suggests that the neural bases of drug craving involve the mesolimbic system but are widespread distributed through the "brain addiction circuit" (**Table 1**). As elucidated by preclinical studies, the accumbens is one of the brain regions that supports drug craving incubation and relapse, together with dorsal striatal and limbic areas (reviewed in [81]). Accordingly, functional neuroimage studies in drug users exposed to drug-associated cues have reported increased activation in either the accumbens, the dorsal striatum, or the limbic regions in correlation with the intensity of craving experienced [82–86].

3. The dorsal striatum: where goal-directed behavior becomes habit

Together with the mesolimbic system, the dorsal striatum is a key brain region to explain addiction. The dorsal striatum, composed by the caudate nucleus and putamen, is a center for sensorimotor integration. It receives excitatory inputs from the thalamus, which is a major relay for sensory signals, and extensive excitatory inputs from cortical areas that are distributed across the striatal subdivisions through the *cortico-striatal circuit* [87] (**Figure 1B**). In this regard, the *dorsomedial striatum* is mainly innervated by cognitive-related prefrontal cortical regions supporting executive functions (and thus it is mostly involved in goal-directed behavioral control), while the *dorsolateral striatum* mostly receives input from primary sensory and motor cortices (and thus seems more involved in habit learning and motor execution) [55, 88]. Furthermore, the so-called *spiraling nigrostriatal circuit* allows functional and bidirectional serial connections among the dorsal striatum and the reward centers including the accumbens and the dopaminergic neurons in the midbrain [55, 88] (**Figure 1B**).

The dorsal striatum is critical to control motor learning, motor planning, and motor execution [87] and to engage in motivated goal-directed behaviors, including those needed for survival [89]. Strikingly, hungry mice with dorsal striatal malfunction will not initiate feeding behavior even when food is placed right in front of them, nor they would explore a novel environment [89]. Considering this, the dorsal striatum is essential for instrumental learning [87, 90], but its function differs from the mesolimbic system's role. While the accumbens predicts the occurrence of a reward in the presence of reward-associated stimuli, the dorsal striatum is in charge of selecting and initiating the actions or movement patterns that are adequate to obtain such expected reward in a certain environment. However, once the reward-associated cue is repeatedly paired with an appropriate action, that results successfully rewarded, the action progressively becomes a routinary response that is automatically elicited by the associated stimulus. In other words, the action becomes a *habit* [55, 91, 92]. Compared to planned goal-directed responses, habits are less flexible and more prone to errors since they are executed unconsciously, based on past performance, without thoughtful evaluation of the current situation. Despite this, habits are highly adaptive for normal everyday functioning, since they allow the dorsal striatum to rapidly select and perform common responses without demanding cognitive and attentional resources that may be directed elsewhere [93].

Nevertheless, when habits involve undesired drug-seeking and drug-taking responses, they entail a core problem in drug addiction. In fact, some authors conceptualize addiction as a "shift" of behavioral control from the accumbens to the dorsal striatal regions as drug-induced neuroplasticity hijack the striatal circuits responsible for habit forming [55, 91, 92] (**Figure 2**). Similarly to what is reported for the accumbens, there is a depletion in the dorsal striatal dopaminergic signaling as evidenced by lower levels of endogenous dopamine [74, 75] and a reduced availability of the dopaminergic D2 receptors [76, 77, 94, 95]. However, in addition to structural plasticity [96], the dorsal striatal neurons may trigger concomitant synaptic changes in the presence of drugs, resulting either in LTP or LTD in response to the midbrain dopaminergic input [91, 92], together with a potentiated glutamatergic transmission attending to an increased density and synaptic facilitation of glutamate receptors [96–98]. Interestingly, while many brain regions in persons with SUDs usually show a reduced gray matter volume, the dorsal striatum has been found either reduced or hypertrophied in psychostimulant-dependent individuals [99–101]. The progressive transition of drug-seeking from a goal-directed behavior to a compulsive habit under striatal control has been elegantly modeled by animal research. At the initial phases of drug self-administration, the expression of this behavior requires the integrity of both the accumbens and the dorsal striatum [102]. But once the animal is extensively trained for drug-seeking, cue-induced drug-seeking is disrupted by interventions affecting the dorsal striatal region selectively (revised in [55]). Furthermore, animals with extended history of drug exposure will not cease drug-seeking even when this behavior is no longer "rationally" worth it (e.g., when they must endure highly aversive stimuli such as electric shocks to obtain the drug [103]), mimicking habitual drug use despite of negative consequences as found in SUD patients.

In conclusion, striatal neuroplasticity supports the progressive transformation of conscious and voluntary (i.e., goal-directed) drug-taking actions into habits (**Table 1**). Habits are an important cause of relapse as they are compulsive and uncontrollable by the individual and automatically elicited by drug-associated cues. This explains that drug-related stimuli (e.g., an alcohol bottle, a razor blade, a place where the drug was usually consumed, a drug-using companion, or even drug-associated emotions and thoughts) would trigger drug use—usually accompanied by intense craving feelings—despite of efforts to remain abstinent [55, 91, 92] (**Figure 1A**).

4. Limbic regions: the prefrontal cortex, the amygdala, and the hippocampus

4.1 Controlling behavior, memory, and mood

While the addiction theories have traditionally focused on the interaction among the mesolimbic system and the dorsal striatal regions, the limbic brain regions—such as the prefrontal cortex, the hippocampus, and the amygdala—have gained increased attention in addiction [27, 104, 105]. As exposed before, the prefrontal cortex, the hippocampus, and the amygdala are mainly glutamatergic structures all reciprocally interconnected [33] that are closely integrated into the reward brain circuit by receiving direct dopaminergic inputs from the VTA and, conversely, by regulating accumbens and VTA activity (Section 2.1; **Figure 1B**).

The limbic system is classically defined as the brain substrate of "emotion" [106]. The limbic regions modulate the stress response, which is generally stimulated by the amygdala but suppressed by the hippocampus and the prefrontal cortex by inhibitory feedback mechanisms [107]. The amygdala also plays a pivotal role in triggering "unpleasant" emotions and responses such as anxiety and fear [108, 109], though it is also involved in positive emotions and it is activated after either appetitive or aversive stimuli, to evaluate their motivational value [110]. The limbic regions also hold cognitive functions. The prefrontal cortex has a key role in behavioral control, by guiding the dorsal striatum to select appropriate actions through the abovementioned cortico-striatal circuit [88] (Section 3) and by inhibiting or updating inappropriate behaviors (reviewed in [15]). Accordingly, the prefrontal cortex is responsible of higher cognitive process such as planning, reasoning, behavioral flexibility, or decision-making (executive functions), and it holds the "working-memory" capacity that allows to manipulate information that is stored in the short-term (reviewed in [15]). The hippocampus is involved in the acquisition, long-term storage, and further processing (extinction, retrieval, updating, etc.) of declarative memory [111]. Declarative memory includes the semantic memory (verbal information, facts, and concepts), the episodic memory (life events), as well as the spatial memory (contexts and places), so a loss of hippocampal function drives severe anterograde amnesia [111]. Moreover, the hippocampus is important for integrating events that are separated in time or space (thus being crucial for associative learning [112]), and it participates in novelty detection that contributes to recognize previously presented stimuli, allowing to lead exploration and/or cognitive resources to the novel ones [113]. Regarding the amygdala, this region also holds a role in cognition, such as in fear memories [114] or in facilitating the emotional modulation of declarative memory, since emotionally arousing experiences are more strongly consolidated and remembered than neutral ones [115] (**Table 1**).

The initial experiences with drugs would use the regular learning mechanisms in the limbic regions to be acquired and stored in memory [116]. In this way, the prefrontal cortex guides the dorsal striatum and acts as an "ON/OFF switch" for drug-seeking, deciding when this behavior should be allowed or inhibited [105]. Regarding the hippocampus and the amygdala, they interact with the prefrontal cortex and the accumbens for the learning of drug-stimuli associations; and these limbic regions collaborate for the subsequent retrieval, extinction, or reinstatement of the drug-related memories (being the reinstatement, a form of "relapse," that in preclinical models is elicited by drug-associated cues, by stress, or by a low dose of the drug—*priming*) (reviewed in [15, 27, 116]). Since the drug-related experiences are rewarding and emotionally arousing, they activate neurobiological pathways involved in the emotional enhancement of associative memory, which may potentiate their acquisition and subsequent long-term maintenance [116, 117].

4.2 Affective and cognitive alterations are concomitant to drug addiction

After repeated drug exposure, the limbic regions are highly vulnerable to undergo neuroplastic and/or neurodegenerative changes (**Figure 2**). A reduced gray matter volume is often found in the prefrontal cortex, hippocampus, and amygdala of chronic drug users [7, 10, 12, 59, 118], together with a dysregulated expression of genes including those involved in GABA and glutamate neurotransmission

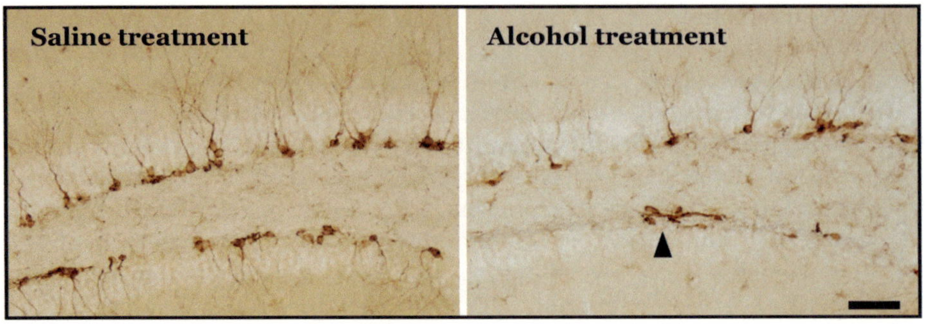

Figure 3.
Reduced adult hippocampal neurogenesis as an example of drug-induced neuroplasticity. Photographs show the hippocampus (dentate gyrus) of mice treated either with saline or ethanol for 8 days (protocol published in our previous work [22]). Young neurons expressing the immature neuron marker doublecortin were stained by immunohistochemistry. Arrow points young neurons showing horizontally disposed nuclei and underdeveloped dendritic tree in the ethanol-treated animal. Scale bar: 100 μm.

[119, 120] and alteration in LTP or LTD processes [121–124]. Particularly, alcohol is associated with severe brain damage and neurotoxicity in the limbic system [12], and sufficient exposure may precipitate severe neurocognitive syndromes such as lasting dementia [125]. Other limbic neuroadaptations induced by addictive drugs involve a reduction of adult hippocampal neurogenesis, as evidenced by a recent *postmortem* study in persons that abused alcohol [126]. The generation, maturation, and functional integration of new neurons in the adult brain—where the dentate gyrus of the hippocampus is a main neurogenic niche—has been extensively described in rodents, for which the new hippocampal neurons participate in many forms of hippocampal-dependent learning and emotional regulation [127]. While the existence and functional implications of adult hippocampal neurogenesis in humans still generate controversy [128], there is currently a wide preclinical evidence supporting that drugs of abuse modulate—mainly reduce—the adult-born hippocampal neurons (**Figure 3**), which has raised interest on the potential involvement of this neuroplastic phenomenon in addiction [27, 116, 129, 130].

Damage of the limbic regions generates the "cognitive" symptoms in drug addiction. The drug-induced neuroplasticity in prefrontal areas involved in the cortico-striatal circuit contributes to the "loss of control" over drug-seeking behavior that becomes further governed by the dorsal striatal habits [105, 131, 132] (Section 3; **Table 1**). The prefrontal "disinhibition" may affect other behavioral domains, promoting impulsivity, impaired decision-making, and more involvement in risky behaviors [133] which, in turn, may contribute to further engagement in drug use (**Figure 1A**). Since the limbic regions are required for associative memory, memories for drug-stimuli associations may become engrained in addiction, being resistant to extinction and forgetting but prone to reinstatement [117, 134, 135]. Therefore, a potentiated function of the limbic regions at the initial experiences with drugs may facilitate their storage in memory; but their impoverished function after repeated drug exposure may impede these memories to be subsequently extinguished. As explained before (Sections 2 and 3), the memories for drug-stimuli associations are relevant in addiction, since they trigger drug craving and habitual drug use responses.

Furthermore, limbic system malfunction in addiction yields a variable degree of cognitive decline that may affect both prefrontal- and hippocampal-dependent domains, including attention, working memory, declarative memory, and executive functions, as evidenced in both drug-exposed animals and in persons with

SUDs (**Table 1**; reviewed in [15]). Cognitive impairment may last for months or years after ceasing drug use, and, in the most severe cases, it may be irreversible (e.g., [18, 125, 136–138]). The cognitive decline has relevant clinical implications, since it is a consistent predictor of addiction treatment dropout and relapse (reviewed in [15]). In this way, it is possible for cognitive impairment to act as an indirect indicator of the extent of malfunction of the limbic regions that are implicated in key behavioral processes that lead to drug use such as behavioral disinhibition or drug craving (**Table 1**). Another possibility is that cognitive impairment may directly compromise the follow-up of addiction treatments by burdening the acquisition of new adaptive information, such as the contents of behavioral therapies that usually require a considerable cognitive effort to be apprehended [139].

Finally, at the emotional level, malfunction of limbic regions during drug withdrawal may curse with a "negative affect" involving stress and anxiety in addition to "loss of reward" (Section 2; **Table 1**) that may trigger drug use by negative reinforcement (i.e., using drugs to escape the aversive emotional state) [78, 140] (**Figure 1A**). In fact, the stress response is frequently dysregulated in persons with SUDs [141] that are vulnerable to stressful experiences, which are a powerful cause of relapse in drug use [81, 142]. Furthermore, SUDs have a high psychiatric comorbidity (~40%) with mood and anxiety disorders [143–145]. Dual pathology complicates the treatment of drug addiction, since an integrative therapeutic approach that involves both the SUD and the comorbid psychiatric disorder may be necessary for these patients [146].

5. Conclusion

This chapter shows that addiction compromises widespread brain neuroplasticity and function, which includes—but is not limited to—key brain regions involved in learning, reward, and motivated behavior. As consequence of repeated drug exposure, probably acting in combination with pre-existing neurobiological vulnerability traits, these regions corrupt their "normal" activity and promote dysfunctional behavior that underlies the etiology and maintenance of the drug addiction disorder. Considering this, therapies directed to promote adaptive neuroplasticity that allows these brain regions to regain their original function are valuable in drug addiction. Importantly, these strategies are not limited to biomedical interventions, but they may include a wide range of behavioral approaches, such as cognitive stimulation, considering that engagement in new and appealing experiences may sculpt brain neuroplasticity, even in the presence of drugs [15]. Therefore, while addiction may be, in a way, a "brain disease," many factors should be taken into account, considering that thoughts, emotions, social, and environmental stimuli ultimately impact the brain.

Acknowledgements

This study was funded by grants from the Spanish Ministry of Economy and Competitiveness (MINECO, Agencia Estatal de Investigación) co-founded by the European Research Development Fund—AEI/FEDER, UE—("Jóvenes Investigadores" grant PSI2015-73156-JIN to E.C.O.; grant PSI2017-82604R to L.J.S and "Juan de la Cierva-Formación" grant FJCI-2015-23925 to P.S.P.) and by the University of Malaga (Plan Propio 2017—PPIT.UMA.B1.2017/38 to P.S.P).

GOBIERNO DE ESPAÑA
MINISTERIO DE ECONOMÍA, INDUSTRIA Y COMPETITIVIDAD
AGENCIA ESTATAL DE INVESTIGACIÓN
EUROPEAN UNION
EUROPEAN REGIONAL DEVELOPMENT FUND
A WAY TO MAKE EUROPE

Conflict of interest

Authors declare no conflicts of interest.

List of abbreviations and acronyms

Acb	accumbens
ACTH	acetylcholine
BLA	basolateral amygdala
D2 receptor	dopamine receptor “D2”
Dstr	dorsal striatum
GABA	γ-aminobutyric acid
Hipp	hippocampus
LTP	long-term potentiation
LTD	long-term depression
PFC	prefrontal cortex
Sep	septum
SMC	sensorimotor cortex
SUD	substance use disorder
VP	ventral pallidum
VTA	ventral tegmental area

Author details

Patricia Sampedro-Piquero[1,2*], Luis J. Santín[1,2*] and Estela Castilla-Ortega[1,3*]

1 Instituto de Investigación Biomédica de Málaga-IBIMA, Spain

2 Departamento de Psicobiología y Metodología de las Ciencias del Comportamiento, Facultad de Psicología, Universidad de Málaga, Spain

3 Unidad de Gestión Clínica de Salud Mental, Hospital Regional Universitario de Málaga, Spain

*Address all correspondence to: patricia.sampedro@uma.es; luis@uma.es and estela.castilla@ibima.eu

References

[1] GBD 2016 Alcohol Collaborators. Alcohol use and burden for 195 countries and territories, 1990-2016: A systematic analysis for the Global Burden of Disease Study 2016. Lancet. 2018;**392**:1015-1035

[2] United Nations Office on Drugs and Crime. World Drug Report 2018: United Nations publication, Sales No. E.18.XI.9; 2018

[3] Marlatt GA, Baer JS. Addictive behaviors: Etiology and treatment. Annual Review of Psychology. 1988;**39**:223-252

[4] Volkow ND, Koob GF, McLellan AT. Neurobiologic advances from the brain disease model of addiction. The New England Journal of Medicine. 2016;**374**:363-371

[5] Hammer R, Dingel M, Ostergren J, Partridge B, McCormick J, Koenig BA. Addiction: Current criticism of the brain disease paradigm. AJOB Neuroscience. 2013;**4**:27-32

[6] Levy N. Addiction is not a brain disease (and it matters). Frontiers in Psychiatry. 2013;**4**:24

[7] Battistella G, Fornari E, Annoni JM, Chtioui H, Dao K, Fabritius M, et al. Long-term effects of cannabis on brain structure. Neuropsychopharmacology. 2014;**39**:2041-2048

[8] Connolly CG, Bell RP, Foxe JJ, Garavan H. Dissociated grey matter changes with prolonged addiction and extended abstinence in cocaine users. PLoS One. 2013;**8**:e59645

[9] Ersche KD, Jones PS, Williams GB, Turton AJ, Robbins TW, Bullmore ET. Abnormal brain structure implicated in stimulant drug addiction. Science. 2012;**335**:601-604

[10] Gallinat J, Meisenzahl E, Jacobsen LK, Kalus P, Bierbrauer J, Kienast T, et al. Smoking and structural brain deficits: A volumetric MR investigation. The European Journal of Neuroscience. 2006;**24**:1744-1750

[11] Hall MG, Alhassoon OM, Stern MJ, Wollman SC, Kimmel CL, Perez-Figueroa A, et al. Gray matter abnormalities in cocaine versus methamphetamine-dependent patients: A neuroimaging meta-analysis. The American Journal of Drug and Alcohol Abuse. 2015;**41**:290-299

[12] Kril JJ, Halliday GM. Brain shrinkage in alcoholics: A decade on and what have we learned? Progress in Neurobiology. 1999;**58**:381-387

[13] Li M, Tian J, Zhang R, Qiu Y, Wen X, Ma X, et al. Abnormal cortical thickness in heroin-dependent individuals. NeuroImage. 2014;**88**:295-307

[14] Sutherland MT, McHugh MJ, Pariyadath V, Stein EA. Resting state functional connectivity in addiction: Lessons learned and a road ahead. NeuroImage. 2012;**62**:2281-2295

[15] Sampedro-Piquero P, de Guevara-Miranda DL, Pavon FJ, Serrano A, Suarez J, de Fonseca FR, et al. Neuroplastic and cognitive impairment in substance use disorders: A therapeutic potential of cognitive stimulation. Neuroscience & Biobehavioral Reviews. 2018. (In Press) DOI: 10.1016/j.neubiorev.2018.11.015

[16] Ersche KD, Williams GB, Robbins TW, Bullmore ET. Meta-analysis of structural brain abnormalities associated with stimulant drug dependence and neuroimaging of addiction vulnerability and resilience. Current Opinion in Neurobiology. 2013;**23**:615-624

[17] Homberg JR, Karel P, Verheij MM. Individual differences in cocaine addiction: Maladaptive behavioural traits. Addiction Biology. 2014;**19**:517-528

[18] Vonmoos M, Hulka LM, Preller KH, Minder F, Baumgartner MR, Quednow BB. Cognitive impairment in cocaine users is drug-induced but partially reversible: Evidence from a longitudinal study. Neuropsychopharmacology. 2014;**39**:2200-2210

[19] Belin D, Berson N, Balado E, Piazza PV, Deroche-Gamonet V. High-novelty-preference rats are predisposed to compulsive cocaine self-administration. Neuropsychopharmacology. 2011;**36**:569-579

[20] Jupp B, Caprioli D, Dalley JW. Highly impulsive rats: Modelling an endophenotype to determine the neurobiological, genetic and environmental mechanisms of addiction. Disease Models & Mechanisms. 2013;**6**:302-311

[21] Ladron de Guevara-Miranda D, Millon C, Rosell-Valle C, Perez-Fernandez M, Missiroli M, Serrano A, et al. Long-lasting memory deficits in mice withdrawn from cocaine are concomitant with neuroadaptations in hippocampal basal activity, GABAergic interneurons and adult neurogenesis. Disease Models & Mechanisms. 2017;**10**:323-336

[22] Sanchez-Marin L, Ladron de Guevara-Miranda D, Manas-Padilla MC, Alen F, Moreno-Fernandez RD, Diaz-Navarro C, et al. Systemic blockade of LPA1/3 lysophosphatidic acid receptors by ki16425 modulates the effects of ethanol on the brain and behavior. Neuropharmacology. 2018;**133**:189-201

[23] Sanchez-Marin L, Pavon FJ, Decara J, Suarez J, Gavito A, Castilla-Ortega E, et al. Effects of intermittent alcohol exposure on emotion and cognition: A potential role for the endogenous cannabinoid system and neuroinflammation. Frontiers in Behavioral Neuroscience. 2017;**11**:15

[24] Berridge KC, Robinson TE. What is the role of dopamine in reward: Hedonic impact, reward learning, or incentive salience? Brain Research. Brain Research Reviews. 1998;**28**:309-369

[25] Ikemoto S. Dopamine reward circuitry: Two projection systems from the ventral midbrain to the nucleus accumbens-olfactory tubercle complex. Brain Research Reviews. 2007;**56**:27-78

[26] Kauer JA. Learning mechanisms in addiction: Synaptic plasticity in the ventral tegmental area as a result of exposure to drugs of abuse. Annual Review of Physiology. 2004;**66**:447-475

[27] Castilla-Ortega E, Serrano A, Blanco E, Araos P, Suarez J, Pavon FJ, et al. A place for the hippocampus in the cocaine addiction circuit: Potential roles for adult hippocampal neurogenesis. Neuroscience and Biobehavioral Reviews. 2016;**66**:15-32

[28] Floresco SB, Todd CL, Grace AA. Glutamatergic afferents from the hippocampus to the nucleus accumbens regulate activity of ventral tegmental area dopamine neurons. The Journal of Neuroscience. 2001;**21**:4915-4922

[29] Rahman S, McBride WJ. Feedback control of mesolimbic somatodendritic dopamine release in rat brain. Journal of Neurochemistry. 2000;**74**:684-692

[30] Faget L, Osakada F, Duan J, Ressler R, Johnson AB, Proudfoot JA, et al. Afferent inputs to neurotransmitter-defined cell types in the ventral tegmental area. Cell Reports. 2016;**15**:2796-2808

[31] Haber SN, Fudge JL, McFarland NR. Striatonigrostriatal pathways in primates form an ascending spiral

from the shell to the dorsolateral striatum. The Journal of Neuroscience. 2000;**20**:2369-2382

[32] Geisler S, Wise RA. Functional implications of glutamatergic projections to the ventral tegmental area. Reviews in the Neurosciences. 2008;**19**:227-244

[33] Jin J, Maren S. Prefrontal-hippocampal interactions in memory and emotion. Frontiers in Systems Neuroscience. 2015;**9**:170

[34] Schultz W, Dayan P, Montague PR. A neural substrate of prediction and reward. Science. 1997;**275**:1593-1599

[35] Bunzeck N, Doeller CF, Dolan RJ, Duzel E. Contextual interaction between novelty and reward processing within the mesolimbic system. Human Brain Mapping. 2012;**33**:1309-1324

[36] Menon V, Levitin DJ. The rewards of music listening: Response and physiological connectivity of the mesolimbic system. NeuroImage. 2005;**28**:175-184

[37] Cooper JC, Knutson B. Valence and salience contribute to nucleus accumbens activation. NeuroImage. 2008;**39**:538-547

[38] Rinaldi A, Oliverio A, Mele A. Spatial memory, plasticity and nucleus accumbens. Reviews in the Neurosciences. 2012;**23**:527-541

[39] Nelson AJ, Thur KE, Marsden CA, Cassaday HJ. Dissociable roles of dopamine within the core and medial shell of the nucleus accumbens in memory for objects and place. Behavioral Neuroscience. 2010;**124**:789-799

[40] Fadok JP, Darvas M, Dickerson TM, Palmiter RD. Long-term memory for pavlovian fear conditioning requires dopamine in the nucleus accumbens and basolateral amygdala. PLoS One. 2010;**5**:e12751

[41] Hernandez PJ, Sadeghian K, Kelley AE. Early consolidation of instrumental learning requires protein synthesis in the nucleus accumbens. Nature Neuroscience. 2002;**5**:1327-1331

[42] Alcaro A, Huber R, Panksepp J. Behavioral functions of the mesolimbic dopaminergic system: An affective neuroethological perspective. Brain Research Reviews. 2007;**56**:283-321

[43] Day JJ, Carelli RM. The nucleus accumbens and Pavlovian reward learning. The Neuroscientist. 2007;**13**:148-159

[44] Bennett BA, Hollingsworth CK, Martin RS, Harp JJ. Methamphetamine-induced alterations in dopamine transporter function. Brain Research. 1998;**782**:219-227

[45] Brodie MS, Dunwiddie TV. Cocaine effects in the ventral tegmental area: Evidence for an indirect dopaminergic mechanism of action. Naunyn-Schmiedeberg's Archives of Pharmacology. 1990;**342**:660-665

[46] Brodie MS, Pesold C, Appel SB. Ethanol directly excites dopaminergic ventral tegmental area reward neurons. Alcoholism, Clinical and Experimental Research. 1999;**23**:1848-1852

[47] French ED. delta9-Tetrahydrocannabinol excites rat VTA dopamine neurons through activation of cannabinoid CB1 but not opioid receptors. Neuroscience Letters. 1997;**226**:159-162

[48] Liu L, Zhao-Shea R, McIntosh JM, Gardner PD, Tapper AR. Nicotine persistently activates ventral tegmental area dopaminergic neurons via nicotinic acetylcholine receptors containing

alpha4 and alpha6 subunits. Molecular Pharmacology. 2012;**81**:541-548

[49] Uramura K, Yada T, Muroya S, Shioda S, Shiratani T, Takigawa M. Methamphetamine induces cytosolic Ca^{2+} oscillations in the VTA dopamine neurons. Neuroreport. 2000;**11**:1057-1061

[50] Johnson SW, North RA. Opioids excite dopamine neurons by hyperpolarization of local interneurons. The Journal of Neuroscience. 1992;**12**:483-488

[51] Steffensen SC, Taylor SR, Horton ML, Barber EN, Lyle LT, Stobbs SH, et al. Cocaine disinhibits dopamine neurons in the ventral tegmental area via use-dependent blockade of GABA neuron voltage-sensitive sodium channels. The European Journal of Neuroscience. 2008;**28**:2028-2040

[52] Drevets WC, Gautier C, Price JC, Kupfer DJ, Kinahan PE, Grace AA, et al. Amphetamine-induced dopamine release in human ventral striatum correlates with euphoria. Biological Psychiatry. 2001;**49**:81-96

[53] Boye SM, Grant RJ, Clarke PB. Disruption of dopaminergic neurotransmission in nucleus accumbens core inhibits the locomotor stimulant effects of nicotine and D-amphetamine in rats. Neuropharmacology. 2001;**40**:792-805

[54] Sellings LH, McQuade LE, Clarke PB. Evidence for multiple sites within rat ventral striatum mediating cocaine-conditioned place preference and locomotor activation. The Journal of Pharmacology and Experimental Therapeutics. 2006;**317**:1178-1187

[55] Everitt BJ. Neural and psychological mechanisms underlying compulsive drug seeking habits and drug memories—Indications for novel treatments of addiction. The European Journal of Neuroscience. 2014;**40**:2163-2182

[56] Corrigall WA, Franklin KB, Coen KM, Clarke PB. The mesolimbic dopaminergic system is implicated in the reinforcing effects of nicotine. Psychopharmacology (Berl). 1992;**107**:285-289

[57] Pettit HO, Ettenberg A, Bloom FE, Koob GF. Destruction of dopamine in the nucleus accumbens selectively attenuates cocaine but not heroin self-administration in rats. Psychopharmacology. 1984;**84**:167-173

[58] Spyraki C, Fibiger HC, Phillips AG. Attenuation of heroin reward in rats by disruption of the mesolimbic dopamine system. Psychopharmacology. 1983;**79**:278-283

[59] Makris N, Oscar-Berman M, Jaffin SK, Hodge SM, Kennedy DN, Caviness VS, et al. Decreased volume of the brain reward system in alcoholism. Biological Psychiatry. 2008;**64**:192-202

[60] Schuch-Goi SB, Goi PD, Bermudez M, Fara LS, Kessler FP, Pechansky F, et al. Accumbens volumes are reduced among crack-cocaine users. Neuroscience Letters. 2017;**645**:86-89

[61] Little KY, Ramssen E, Welchko R, Volberg V, Roland CJ, Cassin B. Decreased brain dopamine cell numbers in human cocaine users. Psychiatry Research. 2009;**168**:173-180

[62] Peterson VL, McCool BA, Hamilton DA. Effects of ethanol exposure and withdrawal on dendritic morphology and spine density in the nucleus accumbens core and shell. Brain Research. 2015;**1594**:125-135

[63] Robinson TE, Kolb B. Structural plasticity associated with exposure to drugs of abuse. Neuropharmacology. 2004;**47**(Suppl 1):33-46

[64] Spiga S, Lintas A, Migliore M, Diana M. Altered architecture and functional consequences of the mesolimbic dopamine system in cannabis dependence. Addiction Biology. 2010;**15**:266-276

[65] Collo G, Cavalleri L, Spano P. Structural plasticity in mesencephalic dopaminergic neurons produced by drugs of abuse: Critical role of BDNF and dopamine. Frontiers in Pharmacology. 2014;**5**:259

[66] Sklair-Tavron L, Shi WX, Lane SB, Harris HW, Bunney BS, Nestler EJ. Chronic morphine induces visible changes in the morphology of mesolimbic dopamine neurons. Proceedings of the National Academy of Sciences of the United States of America. 1996;**93**:11202-11207

[67] Argilli E, Sibley DR, Malenka RC, England PM, Bonci A. Mechanism and time course of cocaine-induced long-term potentiation in the ventral tegmental area. The Journal of Neuroscience. 2008;**28**:9092-9100

[68] Shen H, Kalivas PW. Reduced LTP and LTD in prefrontal cortex synapses in the nucleus accumbens after heroin self-administration. The International Journal of Neuropsychopharmacology. 2013;**16**:1165-1167

[69] Self DW. Regulation of drug-taking and -seeking behaviors by neuroadaptations in the mesolimbic dopamine system. Neuropharmacology. 2004;**47**(Suppl 1):242-255

[70] Steketee JD, Kalivas PW. Drug wanting: Behavioral sensitization and relapse to drug-seeking behavior. Pharmacological Reviews. 2011;**63**:348-365

[71] Trujillo KA, Kubota KS, Warmoth KP. Continuous administration of opioids produces locomotor sensitization. Pharmacology, Biochemistry, and Behavior. 2004;**79**:661-669

[72] Leyton M. Conditioned and sensitized responses to stimulant drugs in humans. Progress in Neuro-Psychopharmacology & Biological Psychiatry. 2007;**31**:1601-1613

[73] Volkow ND, Wang GJ, Fowler JS, Tomasi D, Telang F, Baler R. Addiction: Decreased reward sensitivity and increased expectation sensitivity conspire to overwhelm the brain's control circuit. BioEssays. 2010;**32**:748-755

[74] Kish SJ, Kalasinsky KS, Derkach P, Schmunk GA, Guttman M, Ang L, et al. Striatal dopaminergic and serotonergic markers in human heroin users. Neuropsychopharmacology. 2001;**24**:561-567

[75] Martinez D, Greene K, Broft A, Kumar D, Liu F, Narendran R, et al. Lower level of endogenous dopamine in patients with cocaine dependence: Findings from PET imaging of D(2)/D(3) receptors following acute dopamine depletion. The American Journal of Psychiatry. 2009;**166**:1170-1177

[76] Martinez D, Saccone PA, Liu F, Slifstein M, Orlowska D, Grassetti A, et al. Deficits in dopamine D(2) receptors and presynaptic dopamine in heroin dependence: Commonalities and differences with other types of addiction. Biological Psychiatry. 2012;**71**:192-198

[77] Volkow ND, Wang GJ, Fowler JS, Logan J, Hitzemann R, Ding YS, et al. Decreases in dopamine receptors but not in dopamine transporters in alcoholics. Alcoholism, Clinical and Experimental Research. 1996;**20**:1594-1598

[78] Koob GF. Addiction is a reward deficit and stress surfeit disorder. Frontiers in Psychiatry. 2013;**4**:72

[79] Ahmed SH. Imbalance between drug and non-drug reward availability: A major risk factor for addiction. European Journal of Pharmacology. 2005;**526**:9-20

[80] Berridge KC, Robinson TE. Liking, wanting, and the incentive-sensitization theory of addiction. The American Psychologist. 2016;**71**:670-679

[81] Bossert JM, Marchant NJ, Calu DJ, Shaham Y. The reinstatement model of drug relapse: Recent neurobiological findings, emerging research topics, and translational research. Psychopharmacology. 2013;**229**:453-476

[82] Childress AR, Mozley PD, McElgin W, Fitzgerald J, Reivich M, O'Brien CP. Limbic activation during cue-induced cocaine craving. The American Journal of Psychiatry. 1999;**156**:11-18

[83] Fotros A, Casey KF, Larcher K, Verhaeghe JA, Cox SM, Gravel P, et al. Cocaine cue-induced dopamine release in amygdala and hippocampus: A high-resolution PET [(1)(8)F]fallypride study in cocaine dependent participants. Neuropsychopharmacology. 2013;**38**:1780-1788

[84] Garavan H, Pankiewicz J, Bloom A, Cho JK, Sperry L, Ross TJ, et al. Cue-induced cocaine craving: Neuroanatomical specificity for drug users and drug stimuli. The American Journal of Psychiatry. 2000;**157**:1789-1798

[85] Huang Y, Mohan A, De Ridder D, Sunaert S, Vanneste S. The neural correlates of the unified percept of alcohol-related craving: A fMRI and EEG study. Scientific Reports. 2018;**8**:923

[86] Li Q, Wang Y, Zhang Y, Li W, Yang W, Zhu J, et al. Craving correlates with mesolimbic responses to heroin-related cues in short-term abstinence from heroin: An event-related fMRI study. Brain Research. 2012;**1469**:63-72

[87] Graybiel AM, Aosaki T, Flaherty AW, Kimura M. The basal ganglia and adaptive motor control. Science. 1994;**265**:1826-1831

[88] Haber SN. Corticostriatal circuitry. Dialogues in Clinical Neuroscience. 2016;**18**:7-21

[89] Palmiter RD. Dopamine signaling in the dorsal striatum is essential for motivated behaviors: Lessons from dopamine-deficient mice. Annals of the New York Academy of Sciences. 2008;**1129**:35-46

[90] O'Doherty J, Dayan P, Schultz J, Deichmann R, Friston K, Dolan RJ. Dissociable roles of ventral and dorsal striatum in instrumental conditioning. Science. 2004;**304**:452-454

[91] Belin D, Everitt BJ. Cocaine seeking habits depend upon dopamine-dependent serial connectivity linking the ventral with the dorsal striatum. Neuron. 2008;**57**:432-441

[92] Gerdeman GL, Partridge JG, Lupica CR, Lovinger DM. It could be habit forming: Drugs of abuse and striatal synaptic plasticity. Trends in Neurosciences. 2003;**26**:184-192

[93] Malvaez MW, K.M. Regulation of habit formation in the dorsal striatum. Current Opinion in Behavioral Sciences. 2018;**20**:67-74

[94] Martinez D, Broft A, Foltin RW, Slifstein M, Hwang DR, Huang Y, et al. Cocaine dependence and d2 receptor availability in the functional subdivisions of the striatum: Relationship with cocaine-seeking behavior. Neuropsychopharmacology. 2004;**29**:1190-1202

[95] Volkow ND, Chang L, Wang GJ, Fowler JS, Franceschi D, Sedler M, et al. Loss of dopamine transporters in methamphetamine abusers recovers

with protracted abstinence. The Journal of Neuroscience. 2001;**21**:9414-9418

[96] Jedynak JP, Uslaner JM, Esteban JA, Robinson TE. Methamphetamine-induced structural plasticity in the dorsal striatum. The European Journal of Neuroscience. 2007;**25**:847-853

[97] Cuzon Carlson VC, Seabold GK, Helms CM, Garg N, Odagiri M, Rau AR, et al. Synaptic and morphological neuroadaptations in the putamen associated with long-term, relapsing alcohol drinking in primates. Neuropsychopharmacology. 2011;**36**:2513-2528

[98] Parikh V, Naughton SX, Shi X, Kelley LK, Yegla B, Tallarida CS, et al. Cocaine-induced neuroadaptations in the dorsal striatum: Glutamate dynamics and behavioral sensitization. Neurochemistry International. 2014;**75**:54-65

[99] Barros-Loscertales A, Garavan H, Bustamante JC, Ventura-Campos N, Llopis JJ, Belloch V, et al. Reduced striatal volume in cocaine-dependent patients. NeuroImage. 2011;**56**:1021-1026

[100] Chang L, Alicata D, Ernst T, Volkow N. Structural and metabolic brain changes in the striatum associated with methamphetamine abuse. Addiction. 2007;**102**(Suppl 1):16-32

[101] Jacobsen LK, Giedd JN, Gottschalk C, Kosten TR, Krystal JH. Quantitative morphology of the caudate and putamen in patients with cocaine dependence. The American Journal of Psychiatry. 2001;**158**:486-489

[102] Suto N, Wise RA, Vezina P. Dorsal as well as ventral striatal lesions affect levels of intravenous cocaine and morphine self-administration in rats. Neuroscience Letters. 2011;**493**:29-32

[103] Pelloux Y, Everitt BJ, Dickinson A. Compulsive drug seeking by rats under punishment: Effects of drug taking history. Psychopharmacology. 2007;**194**:127-137

[104] Koob GF. Brain stress systems in the amygdala and addiction. Brain Research. 2009;**1293**:61-75

[105] Peters J, Pattij T, De Vries TJ. Targeting cocaine versus heroin memories: Divergent roles within ventromedial prefrontal cortex. Trends in Pharmacological Sciences. 2013;**34**:689-695

[106] Rajmohan V, Mohandas E. The limbic system. Indian Journal of Psychiatry. 2007;**49**:132-139

[107] Herman JP, McKlveen JM, Solomon MB, Carvalho-Netto E, Myers B. Neural regulation of the stress response: Glucocorticoid feedback mechanisms. Brazilian Journal of Medical and Biological Research. 2012;**45**:292-298

[108] Davis M, Walker DL, Miles L, Grillon C. Phasic vs sustained fear in rats and humans: Role of the extended amygdala in fear vs anxiety. Neuropsychopharmacology. 2010;**35**:105-135

[109] Martijena ID, Molina VA. The influence of stress on fear memory processes. Brazilian Journal of Medical and Biological Research. 2012;**45**:308-313

[110] Zald DH. The human amygdala and the emotional evaluation of sensory stimuli. Brain Research. Brain Research Reviews. 2003;**41**:88-123

[111] Squire LR. Memory and the hippocampus: A synthesis from findings with rats, monkeys, and humans. Psychological Review. 1992;**99**:195-231

[112] Mayes A, Montaldi D, Migo E. Associative memory and the medial temporal lobes. Trends in Cognitive Sciences. 2007;**11**:126-135

[113] Knight R. Contribution of human hippocampal region to novelty detection. Nature. 1996;**383**:256-259

[114] Phillips RG, LeDoux JE. Differential contribution of amygdala and hippocampus to cued and contextual fear conditioning. Behavioral Neuroscience. 1992;**106**:274-285

[115] McGaugh JL. The amygdala modulates the consolidation of memories of emotionally arousing experiences. Annual Review of Neuroscience. 2004;**27**:1-28

[116] Castilla-Ortega E, Ladron de Guevara-Miranda D, Serrano A, Pavon FJ, Suarez J, Rodriguez de Fonseca F, et al. The impact of cocaine on adult hippocampal neurogenesis: Potential neurobiological mechanisms and contributions to maladaptive cognition in cocaine addiction disorder. Biochemical Pharmacology. 2017;**141**:100-117

[117] Torregrossa MM, Corlett PR, Taylor JR. Aberrant learning and memory in addiction. Neurobiology of Learning and Memory. 2011;**96**:609-623

[118] Orikabe L, Yamasue H, Inoue H, Takayanagi Y, Mozue Y, Sudo Y, et al. Reduced amygdala and hippocampal volumes in patients with methamphetamine psychosis. Schizophrenia Research. 2011;**132**:183-189

[119] Enoch MA, Zhou Z, Kimura M, Mash DC, Yuan Q, Goldman D. GABAergic gene expression in postmortem hippocampus from alcoholics and cocaine addicts; corresponding findings in alcohol-naive P and NP rats. PLoS One. 2012;7:e29369

[120] Liu J, Lewohl JM, Harris RA, Iyer VR, Dodd PR, Randall PK, et al. Patterns of gene expression in the frontal cortex discriminate alcoholic from nonalcoholic individuals. Neuropsychopharmacology. 2006;**31**:1574-1582

[121] Fu Y, Pollandt S, Liu J, Krishnan B, Genzer K, Orozco-Cabal L, et al. Long-term potentiation (LTP) in the central amygdala (CeA) is enhanced after prolonged withdrawal from chronic cocaine and requires CRF1 receptors. Journal of Neurophysiology. 2007;**97**:937-941

[122] Huang CC, Lin HJ, Hsu KS. Repeated cocaine administration promotes long-term potentiation induction in rat medial prefrontal cortex. Cerebral Cortex. 2007;**17**:1877-1888

[123] Huang YC, Wang SJ, Chiou LC, Gean PW. Mediation of amphetamine-induced long-term depression of synaptic transmission by CB1 cannabinoid receptors in the rat amygdala. The Journal of Neuroscience. 2003;**23**:10311-10320

[124] Izumi Y, Nagashima K, Murayama K, Zorumski CF. Acute effects of ethanol on hippocampal long-term potentiation and long-term depression are mediated by different mechanisms. Neuroscience. 2005;**136**:509-517

[125] Brust JC. Ethanol and cognition: Indirect effects, neurotoxicity and neuroprotection: A review. International Journal of Environmental Research and Public Health. 2010;7:1540-1557

[126] Le Maitre TW, Dhanabalan G, Bogdanovic N, Alkass K, Druid H. Effects of alcohol abuse on proliferating cells, stem/progenitor cells, and immature neurons in the adult human hippocampus. Neuropsychopharmacology. 2018;**43**:690-699

[127] Castilla-Ortega E, Pedraza C, Estivill-Torrus G, Santin LJ. When is adult hippocampal neurogenesis necessary for learning? Evidence

from animal research. Reviews in the Neurosciences. 2011;**22**:267-283

[128] Sorrells SF, Paredes MF, Cebrian-Silla A, Sandoval K, Qi D, Kelley KW, et al. Human hippocampal neurogenesis drops sharply in children to undetectable levels in adults. Nature. 2018;**555**:377-381

[129] Chambers RA. Adult hippocampal neurogenesis in the pathogenesis of addiction and dual diagnosis disorders. Drug and Alcohol Dependence. 2013;**130**:1-12

[130] Jun H, Mohammed Qasim Hussaini S, Rigby MJ, Jang MH. Functional role of adult hippocampal neurogenesis as a therapeutic strategy for mental disorders. Neural Plasticity. 2012;**2012**:854285

[131] Jentsch JD, Taylor JR. Impulsivity resulting from frontostriatal dysfunction in drug abuse: Implications for the control of behavior by reward-related stimuli. Psychopharmacology (Berl). 1999;**146**:373-390

[132] Renteria R, Baltz ET, Gremel CM. Chronic alcohol exposure disrupts top-down control over basal ganglia action selection to produce habits. Nature Communications. 2018;**9**:211

[133] de Wit H. Impulsivity as a determinant and consequence of drug use: A review of underlying processes. Addiction Biology. 2009;**14**:22-31

[134] Cleva RMG, J.T. Neuroanatomical structures underlying the extinction of drug-seeking behavior. The Open Addiction Journal. 2010;**3**:63-75

[135] Hyman SE. Addiction: A disease of learning and memory. The American Journal of Psychiatry. 2005;**162**:1414-1422

[136] Darke S, Sims J, McDonald S, Wickes W. Cognitive impairment among methadone maintenance patients. Addiction. 2000;**95**:687-695

[137] Dean AC, Groman SM, Morales AM, London ED. An evaluation of the evidence that methamphetamine abuse causes cognitive decline in humans. Neuropsychopharmacology. 2013;**38**:259-274

[138] Shrivastava A, Johnston M, Tsuang M. Cannabis use and cognitive dysfunction. Indian Journal of Psychiatry. 2011;**53**:187-191

[139] Perry CJ, Lawrence AJ. Addiction, cognitive decline and therapy: Seeking ways to escape a vicious cycle. Genes, Brain, and Behavior. 2017;**16**:205-218

[140] Baker TB, Piper ME, McCarthy DE, Majeskie MR, Fiore MC. Addiction motivation reformulated: An affective processing model of negative reinforcement. Psychological Review. 2004;**111**:33-51

[141] Lovallo WR. Cortisol secretion patterns in addiction and addiction risk. International Journal of Psychophysiology. 2006;**59**:195-202

[142] Sinha R. The role of stress in addiction relapse. Current Psychiatry Reports. 2007;**9**:388-395

[143] Garcia-Marchena N, Araos PF, Barrios V, Sanchez-Marin L, Chowen JA, Pedraz M, et al. Plasma chemokines in patients with alcohol use disorders: Association of CCL11 (Eotaxin-1) with psychiatric comorbidity. Frontiers in Psychiatry. 2016;**7**:214

[144] Merikangas KR, Mehta RL, Molnar BE, Walters EE, Swendsen JD, Aguilar-Gaziola S, et al. Comorbidity of substance use disorders with mood and anxiety disorders: Results of the international consortium in psychiatric epidemiology. Addictive Behaviors. 1998;**23**:893-907

[145] Pedraz M, Araos P, Garcia-Marchena N, Serrano A, Romero-Sanchiz P, Suarez J, et al. Sex differences in psychiatric comorbidity and plasma biomarkers for cocaine addiction in abstinent cocaine-addicted subjects in outpatient settings. Frontiers in Psychiatry. 2015;**6**:17

[146] Kelly TM, Daley DC. Integrated treatment of substance use and psychiatric disorders. Social Work in Public Health. 2013;**28**:388-406